D0131101

ERWIN BAUER'S

# PREDATORS
## of
## North America

of

# ERWIN BAUER'S
# PREDATORS
# North America

Photos by Erwin and Peggy Bauer

Published by Outdoor Life Books, New York

Distributed to the trade by Stackpole Books, Harrisburg, Pennsylvania

Copyright © 1988 by Erwin A. Bauer

Published by
  Grolier Books
  800 North Pearl Street
  Latham, NY 12204

Distributed to the trade by
  Stackpole Books
  Cameron and Kelker Streets
  Harrisburg, PA 17105

Produced by Soderstrom Publishing Group Inc.
Book design by Nai Chang
Typography by David E. Seham Associates Inc.
Text face: 10/12 Goudy Old Style

Brief quotations may be used in critical articles and reviews.
For any other reproduction of the book, however, including
electronic, mechanical, photocopying, recording, or other means,
written permission must be obtained from the publisher.

**Library of Congress Cataloging-in-Publication Data**
Bauer, Erwin A.
    [Predators of North America]
    Erwin Bauer's predators of North America / photos by Erwin and
Peggy Bauer.
        p.      cm.
    Includes index.
    ISBN 1-556-54027-2
    1. Predatory animals—North America.    2. Carnivora—North America.
3. Mammals—North America.    I. Title.    II. Title: Predators of
North America.
QL737.C2B38 1988
599′.053—dc19                                            87–26948
                                                            CIP

Manufactured in the United States of America

# CONTENTS

# PREFACE

Compiling this book about predators—America's wild hunters—soon became far more absorbing than I had anticipated. For many years Peggy and I had been studying some of these carnivores through camera viewfinders, often at very close range. Now I was exploring a dimension of natural history that could not be discovered through photography alone.

All of my subjects were both stealthy and bold. Their senses of sight, scent, and hearing were so keen, so efficient, that even those of the keenest-sensed human couldn't begin to match them. Most of the predators in this book can easily outrun, outswim, and outclimb the world's fastest, strongest, most agile human. They can also live in extreme heat and cold where people would perish without modern technology. How long could a human hunter survive if forced to hunt with only his natural talents, his limited physical ability, and his human senses?

Yet despite the success and efficiency of these wild hunters, I couldn't ignore a common thread that ran throughout my research. With only a few exceptions, most of these remarkable mammals are not thriving today. Indeed, several you will meet are destined to become extinct, and that is truly sad.

We are so drastically changing the face of the earth, so rapidly despoiling the world's wilderness, that suitable habitat for all wildlife—especially the meat-eaters—is fading away. The only chance for their long-term survival is for world attitudes to change and for the human population growth to halt. We must soon concede that all species of wild creatures, including those some people dislike, have a place on this planet.

This project has also had its happier side for a husband-and-wife team of hopeless vagabonds. The photography has taken us to scattered, lonely wilderness areas where predators still hunt. It happens that these are the most exquisite, hauntingly beautiful places on our continent. We have watched wolves hunting and raccoons being hunted. We have had glimpses of jaguars and much longer looks at everything from weasels to grizzly bears. What absolutely indelible experiences!

We hope you will find our words and photographs at least half as exciting and enlightening as they were for us to produce them.

Erwin A. Bauer
Teton Village, Wyoming

# PART ONE
# THE WILD CATS

Late on a slate-gray September afternoon high in the Pass Creek area of the Thorofare Wilderness, southeast of Yellowstone National Park, I huddled beneath a rimrock that overlooked a deep, wild valley. I had been hunting elk for a week, unsuccessfully because the weather had been too sunny and much too warm. The elk were still in their summer high country and during the day were retreating to rest in dense conifer thickets, feeding in open meadows only at night. But now the picture was changing.

A cold wind whined upward from the valley floor and carried the season's first wet snowflakes. It also carried the shrill, haunting bugle challenge of a bull elk not far below. Dressed too lightly (my parka lashed to the saddle of my horse a mile away), I shivered and huddled deeper into a crevice of the rimrock. Then came the bugling again, nearer. As I was about to raise my binocular, I spotted silent movement in brittle brush barely 30 feet away. I sat frozen, following the slowly moving form to my left.

The form soon materialized into a cougar. It emerged and then crouched in full view just below me. It seemed to be stalking the same bull elk I had heard.

For several seconds the big cat crouched motionless save for the tip of its tail, which twitched ever so slightly. Then a vagrant breeze carried my scent to the cat. Without seeming to move its body at all, the cat was suddenly staring at me through cold amber eyes. Then, as silently as it had appeared, the cougar vanished. The pounding sound above the wind was my pulse.

That was not the first cougar I had seen in the wild, but it certainly remains the most indelible of the encounters. This was by far my closest view of any wild cat in North America during a half century of hunting, exploring, and photographing. Although seven different species of wild cats inhabit different parts of North America, not one is easy to see. Many a veteran outdoor explorer has wandered "cat country" a lifetime and never seen one.

North American cats deserve their reputation for being among the most mysterious and elusive mammals on earth. All the wild cats are secretive and mostly nocturnal. Both the darkness and texture of their fur make the animals invisible in poor light. And their camouflage coloration makes them difficult to detect in daylight. Gray and brownish cats blend well into surrounding terrain, while spotted cats blend into the diffused sunlight filtering through foliage.

True wild cats occur on all continents except Australia and Antarctica. In North America as elsewhere, they have adapted to such extreme ecological and climatic conditions as deserts and jungles, plains and high mountains, swamps and coniferous forests. For a fuller appreciation of the wild cats, one must understand several facts about their scientific classification and anatomy.

All cats are carnivores of the family Felidae, which have been roaming the planet for about 40 million years. If you examine the roundish skull of any cat, you will see the long, sharp canine teeth—or fangs—which are forward and slightly recurved. The canines are the "killer teeth." The upper and lower molars in the rear of a cat's jaw are *carnassials,* used for cutting and chewing flesh.

The surface of a cat's tongue is covered with sharp-pointed recurved horny papillae designed both for the laceration of meat and for grooming. Another unique characteristic of all felines is that they walk softly on their toes—hence their graceful, seemingly effortless locomotion.

Cats of the world are divided into three subfamilies: the now-extinct sabre-tooth cats, the true cats, and the cheetahs of the Old World. True cats are subdivided further into "large cats" and "small cats," and differences between the two are anatomical as well as behavioral. Large cats have a hyoid-bone structure, or tongue support, that enables them to roar. Small cats cannot roar. Large cats can also purr, but only when exhaling. Small cats can purr when either inhaling or exhaling.

Eye pupils of large cats are round, while those of small cats contract to a vertical slit. The nose of large cats is covered with hair that reaches the front edge of the nose; on small cats the front edge of the nose is hairless. Large cats always feed in a lying position and do not grip prey with their paws. Small cats eat in a crouched position, holding prey to the ground with their paws. Large cats groom themselves much less often and less thoroughly than small ones.

Actually the term "small cat" can be misleading, since the largest species of the small cats, the American cougar (mountain lion), attains the same size as the African leopard, which is a large cat. The jaguar is the only large cat in North America. The other six species are considered small cats. These include, in addition to the cougar, bobcat, lynx, ocelot, margay, and jaguarundi.

All North American cats except the jaguarundi have white spots on the back of the ears. The most common explanation for this is that the spots serve as a "follow-me" signal for kittens at night. Although many scientists believe that the "follow-me theory" is only an interesting guess, they seem to offer no better explanation.

The American cougar, shown on the next page
and on the previous page, is also called mountain lion,
puma, panther, painter, catamount, and—in
Latin America—*leon.* During the day, cougars often rest
in the cool security of mountain cliffs and caves.

# CHAPTER ONE
# THE COUGAR

Survival is most difficult in winter, in the Rocky Mountains, when cougars must follow prey species to lower elevations. There the big cats are the most vulnerable to man.

December 1964. Big Creek, Idaho Wilderness Area, Idaho. For most of the brutally cold day, Maurice Hornocker and Wilbur Wiles followed a pair of Redbone hounds on the trail of a cougar. The pursuit was painfully slow because the tracks crossed some of the most rugged terrain on the continent. In places the snowdrifts were waist deep. Straps of heavy packs cut into shoulders, and thigh muscles began to cramp. At one point Hornocker considered giving up the chase.

But as late afternoon blended into dusk, the men heard the hysterical baying of their hounds. The cat was treed. Aching legs were forgotten and a frantic footrace with darkness began.

The contest came to an abrupt end on top of a sheer cliff. At the very edge of the dropoff a single dead tree cantilevered out over eternity. The hounds pawed furiously at the tree's base, and a large tawny-gold male lion crouched motionless in its top.

"He couldn't possibly be in a worse spot," Wiles mumbled.

Without a word, Hornocker loaded a drug-filled dart into his tranquilizer gun, aimed at the animal's rump, and squeezed. The cougar hardly flinched. The men waited for the tranquilizer to stupefy the animal. In a few minutes they would be able to rope, weigh, and otherwise examine the immobilized but unharmed cat.

"I'm going up," Hornocker finally said.

Daylight was almost gone when he began climbing the fragile dead tree, now shaking ominously in the wind. Hornocker used tree-climbing spikes and had a rope coiled around his shoulder. He moved only a few inches at a time. A misstep or faulty handhold would have been his last. The cougar watched every move with seemingly glazed eyes.

Hornocker inched upward but then sensed that something was wrong. The tranquilizer hadn't taken full effect yet. The lion started to move downward, swished its tail twice, and then jumped to its death.

That night the tired and discouraged biologists and their hounds camped on a steep mountainside. They huddled around a fire that required constant refueling as they munched survival rations. The temperature plunged almost to zero.

The day's loss was one of very few unfortunate events in a grueling study project Hornocker conducted from 1964 to 1970. Three complete winters were spent in the wilderness. There was also much laboratory work and many short expeditions. Both men studied the North American cougar as it had never been studied. They lived in a wilderness and hunted on foot. When their work was over, Hornocker and Wiles knew more about *Felis concolor* than had ever been known.

**RANGE AND REPUTATION.** The extraordinary adaptability of the cougar has been as much a mystery as its character. Records show that the animal once lived in every state but Alaska and Hawaii, although it now occupies only a small fraction of that original range. Cougars still thrive from Alberta and British Columbia southward through most of the western United States, and through Central and South America all the way to Tierra del Fuego, just north of Antarctica. The same large cat that haunts the waterless red desert canyons of Utah and Arizona also lives in tropical jungles, cloud forests, savannas, ev-

This is the close-up view that Maurice Hornocker often saw when he climbed toward treed cougars in his grueling and hazardous studies in the 1960s, which ultimately revealed far more than had ever been known about the cats.

◁ The spectacular canyon country of the American Southwest is one stronghold of cougars, largely because mule deer exist there in good numbers.

The adaptable cougar is as much at home, as shown here, in central American jungles as it is in the Rocky Mountains.

ergreen woodlands, and scattered sub-alpine habitats wherever there is wilderness. The cougar has the widest distribution of any mammal in the Western Hemisphere. Of all cats, only the leopard of Asia and Africa has a greater natural range.

The cougar's vast geographical range as well as the folklore associated with it account for its many common names. Some of the most common are mountain lion, panther, painter, puma, catamount, and, in Latin America, león. Puma is an American Indian name, while catamount is derived from the early New England term, "cat-a-mountain." Today cougar seems to be a more or less standard reference.

Amerigo Vespucci in 1500 was the first European explorer to mention seeing the león during an exploration of Venezuela. Soon after, Christopher Columbus wrote of seeing this cat along the coast of Honduras and Nicaragua. The first to mention cougars in North America was Sir John Hawkins, an Englishman who in 1565 referred to "lions and tigers" in Florida.

In 1610 Captain John Smith wrote of the cougars he found in Virginia: "There be in this wilde country lions, beares, woulves, foxes, muske catts, hares, fle-inge squirrels, and other squirrels." Dutch settlers in what is now New York believed the American lion was identical to the African lion; but they wondered why none of the pelts brought to them in trade by the Indians had manes, and so assumed they were those of lionesses. The Indians claimed that all the males lived in dangerous mountains many miles to the west, and that they were so terrible no redskin in his right mind would try to hunt them.

Pioneer naturalist Ernest Thompson Seton called the cougar "lithe and splendid beasthood," and that is almost an understatement. The animal is so silent and swift afoot that it seems able to dissolve into its environment. It can climb almost anything, including any limbless tree strong enough to hold its weight. All its senses, particularly eyesight, are very well developed. Powerful jaws with canine teeth 1 to 1½ inches long are formidable weapons, while the lion's feet are soft and fur-padded for stealth.

Throughout the centuries, cougars have given humans little cause for alarm. Few of the reported attacks on people could be verified. A cougar on Vancouver Island, British Columbia, did kill a child; and another

Originally ranging from Canada to the southern tip of South America, the cougar is today mostly restricted to high country where it blends well into its habitat. Few people ever see a wild one.

youngster was attacked and killed in the state of Washington. The British Columbia case involved an animal almost blinded by eye cataracts, while the unfortunate boy in Washington may have invited chase and attack by screaming and running. In another instance a Hinton, Alberta, housewife was able to pull a puma off her child by grabbing it by the ears. Virtually all cougar attacks have been on children, as are a great percentage of attacks by carnivores all over the world.

Still, the cougar has been feared, hated, and pursued with almost religious passion as has every large predatory animal. In the New World, only the gray wolf and, more recently, the coyote have inspired as much emotion and controversy. Even that pioneer environmentalist and sportsman Teddy Roosevelt described the mountain lion as a "big horse-killing cat, destroyer of the deer and lord of stealthy murder with a heart craven and cruel." That is typical of what many have written. Thanks to biologist Maurice Hornocker, we now know that the cougar's evil reputation is hardly deserved.

**THE HORNOCKER STUDIES.** In 1964 Maurice Hornocker was 34, a veteran of submarine service in the Navy, and a graduate student in wildlife management at the University of Montana. He had also taken part in the celebrated Craighead studies of grizzly bears in Yellowstone National Park. With the Craighead brothers he had live-trapped, immobilized, radio-collared, and followed the giant bruins across the vast wilderness of northwestern Wyoming. Maurice Hornocker was in such fine physical condition that he could get around easily in any kind of country. A fellow student commented that ordinary grizzly bears were no match for him.

It was almost inevitable that Hornocker select the heretofore unstudied cougar for his post-graduate work. The studies were assisted by the Idaho Fish and Game Department, the University of British Columbia, the American Museum of Natural History, the Boone and Crockett Club, and the New York Zoological Society. Hornocker and lion tracker Wilbur Wiles also had scientific collecting permits from the state of Idaho that allowed them to kill lions, though they tried earnestly never to do so.

The objectives of their study were to observe cougar populations and determine their impact on big game. Lions had long been accused of reducing and even wiping out deer and elk herds. Studying the cats in

A skillful stalker of hoofed big game, the cougar
depends on a final, fast rush to reach its prey.

the wild would be the only way to learn the truth. Unlike other predators such as bears and most big cats, cougars are almost impossible to bait or lure into traps—a fact that made collecting data infinitely more difficult.

Many outdoorsmen spend lifetimes in America's best lion country and never see one, as already mentioned. In the years that followed, Hornocker himself rarely got more than a fleeting glimpse of a free-roving cougar. The solution was to capture, examine, and individually mark as many lions as possible. Recapturing animals over the years, he hoped, would help him accumulate the information he needed.

The Big Creek drainage in central Idaho was selected for the study because wintering herds of big game concentrate there, and so do cougars. The wilderness has long been known as top cougar-hunting country. Most of the area is so rugged, it can only be traversed on foot.

Hornocker's first step was to hire Wilbur Wiles, a professional lion hunter. Wiles owned several tried-and-true "lion dogs," and the men planned to tree cougars for capture and study, and to mark them for identification.

"We spent three full winters from December to May in the mountains and made millions of footprints together," Hornocker recalls. "During those three winters we never saw another human being. Wilbur was always a tireless, enthusiastic, unfailing source of cheer during the worst weather and snowstorms."

Their own survival as well as the success of the lion project depended on careful planning. Well before the first snowfall, several complete mountain camps were established in strategic places throughout the Big Creek drainage. A good supply of food was also cached at each campsite out of reach of bears. The goal was to have dry overnight refuge fairly handy, no matter where a lion trail might lead them. It didn't always work out that way, and the two spent many nights out under the pines.

During the three complete winters in the mountains, Hornocker and Wiles trekked more than 5,000 miles. The team live-captured and tranquilized 46 different cougars, many of them taken over and over again.

For Hornocker, following cougars into the lonely backcountry for several winters was high adventure, but it was also unadulterated hard work. More than once the pair followed a cougar all day, only to have the tracks wiped out completely by a sudden spring

A mother cougar keeps her eye on twin kittens at
the mouth of a Rocky Mountain cave.

rainstorm or covered by a heavy snowfall. Bone-drenching episodes were not unusual and neither were charley horses.

And there were hazards. Hornocker recalls watching the wind topple a rotten tree only moments after he had tranquilized and retrieved a lion from its top. "Then and there I learned not to climb *every* tree a lion can climb." The biologist also recalls the day he found the bloody place where a male cougar had killed and eaten a young cougar, proving that the species is sometimes cannibalistic.

Once or twice hounds were raked by sharp claws, and immobilized cats revived more quickly than anticipated. In retrospect, however, Hornocker feels that the greatest danger lay in climbing unsteady trees and using ropes to rescue overeager hounds trapped on mountain ledges.

## BIRTH AND DEVELOPMENT.

Hornocker and Wiles soon learned that mature male cougars are very solitary, though the animals breed at any time of the year. Young cats of various ages were caught throughout the winter. The men also learned that male cougars immediately after mating desert the females. About three months after that, kittens are whelped in a dry and remote cave den. There may be as many as six young in a litter, but Hornocker found the average to be 2½ kittens at two-year intervals. Young are born blind and helpless, and the female protects and trains them—a task that requires great care and patience.

As soon as the kittens are mature enough, the mother leads them on training trips one at a time. The training may last until the youngsters are larger than their mother, and may even continue after she has had a second litter. Hornocker recalls an outstanding example of maternal devotion.

One winter Hornocker and Wiles captured a mature female that weighed 98 pounds. Before releasing her, they fitted her with a miniaturized radio transmitter so that they could monitor her daily travels and activities from a distance. The female was caring for two 18-month-old kittens, each of which weighed more than she. When captured later the female kitten weighed 105 pounds, and the male 135 pounds!

During radio tracking the mother killed two six-point bull elk (among other animals) to feed her "babies," which were not yet self-sufficient. Each of those two elk was six to seven times the size of the mother cat.

Hornocker's original Big Creek study area had only ten resident cougars, roughly one per 20 square miles. All of the resident adults had established territories, the boundaries marked by mounds of brush or pine needles scraped together and scented with urine.

While large predators such as hyenas, wolves, and African lions live in packs and share food, the shy, solitary cougar must live by its wits. Cougars rarely invade the territories of others, a behavior pattern biologists call *mutual avoidance*. Until a cougar is mature enough to command its own territory, it wanders somewhat aimlessly, unwelcome even among relatives.

## HUNTING.

Hornocker noted that cougars often place their hind feet directly in the tracks made by their front feet, reducing any chance of snapping a twig or making other noises. Cougars hunt skillfully and will do so as often in daylight as in darkness, especially in places remote from human settlement.

Consider that a cougar in the Idaho wilderness, where winters are long and severe, subsists almost entirely on deer and elk in about equal numbers. Hornocker learned that there is nothing clumsy or haphazard about a mountain lion's kill. With one clawed paw grasping the prey's muzzle, the cat normally snaps the neck at the first or second vertebra so that the animal is dead almost on contact after the spring.

But the cougar doesn't always succeed. Mature elk are so large and powerful that the odds are against the cat. One morning in 1967, tracker Wilbur Wiles came upon a dramatic story etched in a fresh snowfall. It was clearly evident that a cat had attacked a small bull elk in a herd of four or five. But instead of an immediate kill, prey and cat had rolled more than 100 feet down a steep mountainside and crashed into a tree. The elk had escaped, and the cat had wandered away. Wiles released his hounds to follow it. The cougar, a female, was tired and soon treed. Although bloodied about the mouth, she appeared unharmed.

Three weeks later and about 14 miles away, Wiles recaptured the same female, but she was thin and barely able to climb. After she was tranquilized, a detailed examination revealed antler punctures in the shoulder and hind legs, a broken jaw, and canine teeth torn completely out of the skull. The dying cougar was suffering so horribly that Wiles destroyed her.

Even during the best of times, cougars may have to travel far with empty stomachs, a condition that occasionally can get them into trouble. Early in 1984, members of a campground work crew northwest of Boulder, Colorado, were baffled when they found a dead cougar covered by the remains of a bull elk whose neck it had broken. The apparently uninjured cougar appeared to have been fatally pinned down. Laboratory tests revealed that both predator and prey had had empty stomachs and had been near starvation. Perhaps the cougar's strength had simply drained away.

Though cougars kill a good many deer and elk,

An agile climber of trees and cliffs, this cougar uses its high perch to watch
the movements of game far below.

Maurice Hornocker found the overall impact on big game to be small. In fact, the presence of cougars may even be beneficial. The best evidence is that elk and mule deer, America's principal western big-game species, have thrived for centuries in cougar country. There is no sound reason to believe this will change. Examinations of carcasses and bones also revealed that the easiest prey—old and young animals—comprised most of the kill. Seventy-five percent of the elk and 62 percent of the mule deer killed by cougars were either less than 1½ years old or more than nine. Virile breeding-age prey animals are least preyed upon, while poor specimens are culled.

In no case were cougars found to be wanton destroyers of large numbers of game animals just for the sake of killing. After a cougar made a kill, whether alone or feeding young, the carcass was usually covered with debris and revisited until completely eaten.

John Long, a taxidermist and lifelong cougar hunter of Sheridan, Wyoming, has been surprised at how often cougars must settle for snowshoe hares or even squirrels. At least in eastern Wyoming, they also try to eat a lot of porcupines. Almost every cougar carcass Long has ever examined has had porcupine quills embedded in it somewhere.

Although they normally rely on larger game in winter, hungry cougars will also pursue snowshoe hares and other small creatures.

In west Texas, whitetail deer become the main prey of cougars. But the predators rarely manage to kill large and healthy bucks.

Fast and sure-footed over even the most rugged terrain, the cougar still depends on surprise and a quick rush to reach its often speedier prey.

Hornocker confirmed one theory about the relationship between cougars and prey. When a cougar made a kill, the other deer or elk cleared out of the area for safer places. Thus, the cougars caused constant shifting about and mixing of the herds, which cuts down on inbreeding and also promotes lighter browsing of range that might otherwise become overbrowsed.

**THE LARGEST COUGAR.** One of the largest cougars of record was also a hungry one. Tom Ladenburg and his wife, who live near Columbia Falls, Montana, returned home from a shopping trip when their dog, a 25-pound border collie, flushed a cougar from beneath their trailer house. In the ensuing melee, the cat grabbed the dog by the head and Ladenburg shot it. The nearly toothless old lion had a skull that measured 9½ inches long, equaling that of the record skull taken by Teddy Roosevelt in Colorado in the early 1900s. That Montana cougar also had a cracked rib cage, ulcerated front teeth, and an empty stomach.

**COUGARS AND MAN.** In March 1986, five-year-old Laura Small was hiking with her sister and parents in the Ronald W. Caspers Wilderness Regional Park, only 60 miles southeast of Los Angeles, when she was suddenly attacked by a cougar. She suffered severe

bites on the head and legs before another hiker, Gregory Ysais, drove it away with a stick. Hunters later shot the 80-pound cat. Although many biologists are convinced that this cat was either an escaped captive animal or a troublesome pet, released into the wild when it became too mean to handle, the incident nonetheless inflamed an old controversy: man and his possessions versus cougars.

The controversy erupted again recently in the oak and chaparral country along the north fork of the Kings River, east of Fresno. Because of a spate of cattle kills, which may or may not have been caused by cougars, officials of the California Department of Fish and Game and of the U.S. Forest Service jointly proposed a complete extermination of all cougars in a 250-square-mile area. That meant killing between 30 and 90 cats, depending on which population estimate one accepts. Not only had the cougars supposedly been killing cattle wholesale, but these officials claimed the cats were also responsible for a sharp decline in black-tail deer over the preceding 25 years. The officials specifically pointed to the high mortality of fawns that had been radio-collared for study.

Indeed, the deer population did fall substantially in that area of California in the 1970s and 1980s. But it is also true that their habitat continues to be systematically overgrazed and ruined by the same live-

This fine male cougar has been brought to bay in dense thicket by a pack of hounds. Coursing with hounds is exciting, but I see little sport in killing a cornered animal.

stock cattlemen claim the cougars are killing. A far better solution to an ugly situation would be to reduce or eliminate livestock grazing on these public lands.

Susan de Treville, who once studied cougars for the California Department of Fish and Game, feels exterminating cougars in one area could well create a vacuum, "disrupting the lion's social structure surrounding the target area and thereby attract into it an even greater density of lions than originally lived there. So the killing would have to spread."

John Seidensticker, who is now assistant mammal curator in the Smithsonian Museum—and who, like Maurice Hornocker, studied lions in the Idaho wilderness area—has another fear about wiping out the lions. "I'm confident," he states, "that removing the lions will only bring about an increase in coyote and bobcat predation. Both of these species would then hunt in areas they would not inhabit if mountain lions were present."

**THE FUTURE.** Today the cougar is protected at least somewhat as a game species in all western states. The animal is hunted, but usually the hunting is carefully regulated. Indiscriminate killing and bounty hunting are mistakes of the past. Rough estimates place the total number of cougars between 4,000 and 7,500 in the western United States, plus a few thousand more in western Canada. Recently an astonishing number of cougar sightings have been reported in the northeastern United States and maritime Canada.

Late at night in the Nova Scotia wilderness a camper hears a loud, chilling caterwaul. At daylight he finds footprints that may or may not have been made by a large feline. A milk truck driver on a back road in upstate New York reports seeing a large brown animal with a long, curving tail bound across a back road. A forester in Maine claims he's seen a tawny animal chase a whitetail deer acorss a meadow. Hundreds of similar reports have convinced some wildlife scientists that the eastern cougar, *Felis concolor cougar*, may still survive in eastern woodlands. This subspecies has been considered extinct for at least a half century.

I suppose it is barely possible. Beginning the day the first Europeans waded ashore at Plymouth Rock, settlers systematically began to cut down the forests that were the cougar's home. They slaughtered turkeys and deer on which the cat depended for food (and of course the cougars themselves) on sight. Today, much of that forest has grown back, covering a good part of the Northeast. Whitetail deer abound, while moose are becoming more and more common throughout New England and as far south as Massachusetts. Many claim that the cougar has returned, but from where?

Cougars have often been blamed for reduced deer populations when the true cause has been habitat destruction, resulting from real estate development and excessive grazing of livestock.

Similar sightings of cougars have recently been reported in the southern Appalachians. Bob Downing, a biologist for the U.S. Fish and Wildlife Service, concluded a five-year project searching for cougars in Georgia and the Carolinas. Downing found no verifiable evidence that the species still survives, but he is still optimistic. He cites a small population recently rediscovered in Manitoba, Canada, where cougars were unknown for decades. The Manitoba cougars may or may not be of the eastern subspecies *Felis concolor cougar*. Downing believes these may be descendants of cougars that retreated westward ahead of early timber cutting in the East, and that they are now filtering back again. Biologist Virginia Fifield, of the Massachusetts Eastern Cougar Survey team, agrees with Downing and believes that a handful of cougars now live in New England west of the Connecticut River. But the evidence is overwhelming that the sole tiny remnant of cougars east of the Mississippi River clings to existence in the deep South.

**THE FLORIDA PANTHER.** On the damp, dark night of November 2, 1984, an unknown motorist drove south on U.S. 41 through the Florida Everglades. Rounding a curve he struck an animal he may have thought was a large dog. Or perhaps he recognized the large tawny cat suddenly illuminated in his headlights. Apparently he did not stop for a closer look.

Shortly after daybreak another driver, Ronald Townsend, was traveling that same road from Miami west to Naples. A few miles outside of Ochopee, he saw first the eyes and then the form of the animal lying beside the road in the beam of his headlights. Townsend braked to a stop and turned back to investigate. What he found was a Florida panther—still another cougar subspecies, *Felis concolor coryi*—crouched in the edge of heavy brush. Somehow it had managed to drag itself from the pavement before collapsing. Townsend couldn't tell if it was dead or alive.

"I've read about panthers and always wondered about them," he said, "so I thought I'd better find help." Townsend drove as fast as he could to the nearest Florida Highway Patrol station, where he started making urgent telephone calls.

By 10 o'clock that morning John Roboski, project leader in charge of south Florida panther research, was on the scene. Also on hand were local sheriff's deputies to control increasing traffic, and a corps of veterinarians with drugs and medical supplies to treat

A female cougar rests briefly near a desert waterhole in the American Southwest. Sheep and deer will come to drink here.

a rare, critically injured animal. One vet, Ned Johnston, a specialist in exotic cats, was helicoptered to the site. They found the cougar alive, but just barely. The animal managed to crawl away from them and swim partway across a small canal before giving up. It almost drowned before being pulled out in a fish net.

What followed must be among the most elaborate attempts to save a wild creature. The panther was darted in an attempt to tranquilize it, though it recovered sufficiently to chase its rescuers back into the canal. About 20 minutes later it was unconscious, and was gingerly lifted into a small boat that would carry it across the canal. Ned Johnston treated the cat for extreme shock and worked to bring its respiration back down to normal. He held the cat's head in his lap as both were helicoptered to the St. Francis Animal Clinic in Naples. En route the panther, a male about 3½ years old and weighing 125 pounds, was nicknamed Big Guy.

The vet swore that the night before he had dreamed he was bass fishing in the Everglades and had seen a panther on the bank. Now the life of one was in his hands. X-rays revealed that both hind legs and a hind foot had sustained compound fractures. The animal was in terrible shape and was immediately flown to a wildlife research facility in Gainesville where complicated surgery would be possible.

Meanwhile a captive-bred panther was located at Camp Kulaqua, a church youth camp at High Springs. This second animal was rushed to Gainesville and preprared for a one-liter blood transfusion. Twelve hours after the injured panther was found, a team of orthopedic surgeons and veterinarians began an operation that included inserting steel plates in the animal's legs. The operation was successful; Big Guy's wounds slowly healed and the cat is alive and healthy as this is written.

Few if any other panthers are as lucky as Big Guy, a symbol for the rescue of a species many thought had become extinct. In the two years after Big Guy was found alive, at least 15 other panthers died in Florida. Nearly all were road kills, mostly on Alligator Alley— the high-speed toll road that connects Naples with Fort Lauderdale—and on State Route 29. Both roads pass through Fakahatchee Strand, Big Cypress National Preserve, and the Everglades areas. These are also prime cougar areas where an estimated 30 individuals—the sum of the earth's population of this subspecies—now cling to an uncertain existence.

**RANGE AND PHYSIQUE.** The Florida subspecies originally ranged across the southeastern states to the Louisiana-Texas border. Once there was a bounty on its hide in Florida, and it was legally hunted until 1958. Now it is a felony to kill what has become Florida's official state animal. The Florida panther is distinctive, differing from other cougars by the cowlick on its back, flecks of white on its shoulders, and the crook in its tail. Charles B. Cory, former curator of Chicago's Field Museum (and for whom the subspecies is named), described the animal in 1896 as "somewhat smaller and more rufous in color than its northern brethren, and its feet are smaller in proportion to the size of the animal. The average male weighs 100 pounds or more; and the average female weighs 10 to 15 pounds less than the male." To this date very little is known about the life history of *Felis concolor coryi*.

Presently somewhere between five and ten panthers live at least part of the time within the boundaries of Everglades National Park. The park, in turn, is only minutes from Florida's Gold Coast area, one of the fastest-growing regions in the United States. To assist in censusing the animals, the Park Service hired Rov McBride of Alpine, Texas, a big-game tracker and hunter of legendary skill.

Using experienced hounds, McBride found panthers living on lonely hammocks in marshy wetlands, the vast sea of grass that inundates a million acres in the south Florida interior. Even more cats were found in an area locally called Hole of the Donut, which was once farmed but has reverted to the wild. No doubt the panthers have been attracted by the sizable herd of small whitetail deer that live in the Hole. The humid, soggy Florida country stands in stunning contrast to the parched Big Bend country of west Texas where McBride has pursued cougars in the past, or to the thin, cool atmosphere of the Idaho Wilderness where Maurice Hornocker began his work on the species.

Yet we still know relatively little about the American cougar, no matter which name it goes by. Elusive and threatened as it is now, I often wonder if we ever will know much about it.

Present in most of the United States at least in ▷
token numbers, the bobcat is the most numerous
and widely distributed of native wild cats.

# CHAPTER TWO
# THE BOBCAT

This bobcat has captured a scaled quail near a deer feeding station in southern Texas. Many creatures come to freeload on the station's free grain, and therefore some become prey.

B obcats have wandered in and out of my life outdoors for many years. I encountered my first bobcat one cold November day, near Big Indian Lake on Michigan's Upper Peninsula while sitting motionless on a deer stand. The cat appeared and disappeared so quickly that moments passed before I realized what I'd seen. My next bobcat materialized high up in Colorado's Uncompahgre National Forest where I was hunting mule deer. This one came walking down a game trail, stopped about 40 feet away, and stared at me in disbelief before slowly walking away. It never looked back. At home in Wyoming I find many bobcat tracks across the winter snow but have not yet seen bobcats there.

Most of my bobcat encounters have been in the brush country of south Texas where the species, *Lynx rufus,* seems to be more abundant than anywhere else in its extensive range. During the fall and winter many ranch owners in Texas feed grain to their whitetail deer at regular feeding stations. The practice also draws javelinas, turkeys, quail, and many other seed-eating birds, all of which in turn attract bobcats. Early one morning near one such feeder, I saw a bobcat capture a scaled quail from a covey, and another morning the cat almost caught a young tom turkey.

I'm absolutely convinced that people, even skilled woodsmen, rarely come upon bobcats in the wild. Almost always it is the other way around. The best way to see a bobcat is to sit down inconspicuously in bobcat country and be very patient.

**A RECLUSIVE NATURE.** The bobcat, perhaps even more than other wild cats, is a master of concealment. Its ability to remain unseen is a behavioral adaption through which it has survived among larger, stronger carnivores for about 40 million years. Were it not for its short, swift initial rush, the bobcat would be unable to capture most prey because otherwise it is simply not fast enough afoot. Bobcats do not leap on prey from trees as is often claimed. They must rely on ambush, crouching down in available cover and then mounting a lightninglike attack on an unsuspecting victim.

The bobcat's reclusive nature serves it in yet another way. Decidedly antisocial, bobcats prefer to avoid other bobcats. Each occupies a territory that is roughly marked by the boundaries of its neighbors on all sides. Any bobcat's survival depends on whether its territory contains an adequate and renewable supply of smaller creatures to feed upon. By avoiding other cats as much as possible—thereby avoiding fights and injury as well as competition for food—each cat can make the most of what his territory has to offer.

Every bobcat seems to devote a good bit of time to marking its own home territory, especially the fringes, with small jets of strong-smelling urine that repels other cats. Females especially may also claim an area by defecating in the same places until small mounds of excrement are obvious to other cats passing by. Such mounds are often the best clue that a bobcat is living in the area. Nonresident or unestablished bobcats will usually turn away from urine scent stations or piles of stool.

**RANGE.** The size of a bobcat's territory varies greatly with geography, weather, and perhaps most of all with the overall density of prey. Today, a few bobcats live in nearly all of the lower 48 states and throughout most of Mexico. Research reveals that southern bob-

In the American Southwest the bobwhite quail is always potential prey for bobcats, especially where the birds do not have sufficient protective cover.

29

cats need much less space per animal than do those in the north.

From 1969 to 1972 biologist Theodore Bailey made a thorough study of bobcats in southeastern Idaho. By capturing cats alive and fitting them with miniature radios, Bailey covered an area of about 250 square miles of mountainous semidesert that contained 17 adults. One of these, a male, occupied a home range of 42 square miles, or almost one-sixth of the study area. Another male managed to survive on less than 3 square miles. The average was about 15 square miles per animal.

Biologist David Lawhead conducted a similar study in central Arizona with somewhat different results. Here the average home territory of a bobcat was only 2.69 square miles. In sections of south Texas bush country that have not been overgrazed by cattle or cleared for agriculture (and where trapping or predator control is not excessive) the density of bobcats may exceed one per square mile.

I regard *Lynx rufus* as a creature strictly of thick cover and rough country, no matter where in America it survives. In regions surrounding the Great Lakes the species is concentrated in evergreen forests, balsam bogs, and swamps or places where the undergrowth is willow and alder. In the South the bobcat lurks most often in overgrown or second-growth forests, dark thickets, canebrakes, and in swamps such as the edge of Okefenokee in Georgia and the Great Dismal Swamp in Virginia. Western cats may keep to river bottoms or rugged canyons where pinyon pines, juniper, and sagebrush provide cover.

**TRAPPED FOR FUR.** Bobcats are fewer today and more sparsely distributed in the northernmost parts of their range than anywhere else. North country is also where bobcats come under greatest pressure from man. Cold climate produces thicker coats and better fur on all furred creatures, including bobcats. Despite the changing styles and fickle vogue for furs, long-haired or so-called "fun furs" such as bobcat are in fairly constant demand. In times when a bobcat pelt brought $500 or more on the raw fur market, trapping pressure became extremely heavy and bobcat numbers have dipped to dangerously low levels in many northern regions.

The situation is ironic because bobcat fur, compared to that of other furbearers, is brittle, tends to

Photos above and right: In winter the faintly spotted coat of the northern bobcat blends well into the dull, gray-and-white habitat.

30

go limp, sheds easily, and is not readily dyed. Until 1973 it was used very little in quality fur garments. Then an international treaty called CITES, for Convention on International Trade for Endangered Species, caused a drastic change. CITES was intended to save such vanishing foreign cats as the ocelot and cheetah, leopard and jaguar, by preventing trade in their pelts. It also caused international furriers to search for a substitute spotted cat. Unfortunately the bobcat was the most readily available. Today's high-priced fur coats sold in Europe and euphemistically labled "lynx cat" or something similar are in reality a dozen or so bobcats, each from Minnesota or Montana.

In some ways the species is its own worst enemy. Its whiskers serve as super-sensitive antennae, and its hairy ears can detect the slightest twitter of ground-nesting birds. The keen, moist nose easily locates field mice in densest cover. All senses are tuned for instant, energetic response—and to keeping the animal unseen and undetected by humans, prey, and by larger carnivores. But the bobcat simply cannot keep out of the steel traps set by man.

During a long lifetime spent largely in Beltrami County, northern Minnesota, Ronn Swedlund trapped everything from pine squirrels and beavers to wolves and black bears. "The dumbest, the easiest to catch," he assured me, "are bobcats. Show me a fresh set of tracks and I'll have a skinned cat for you in a few days."

It is now becoming apparent in the East, as it has in the West, that any coyote clever enough to reach adulthood can detect and avoid leghold traps in its home range, no matter how carefully sited and baited. But bobcats lack such skills, Swedlund assured me, and most seem compelled to test or even play with any strange object they find. Swedlund has trapped cats that, judging from telltale injuries to other paws, had somehow managed to escape from other traps only to be caught again.

While conducting his study in Idaho, Theodore Bailey used specially padded and offset leghold traps to capture and radio-collar bobcats in his area. His traps snapped shut 103 times altogether, but on the legs of only 66 different bobcats. Of those, 29 were captured a second time, often in almost the same place. Some years ago, before the species' pelts became so popular, trappers hated bobcats because they were consistently found in traps baited for more valuable animals.

**THE PHYSICAL SPECIMEN.** Before daybreak one morning on opening day of the deer season in LaSalle County, Texas, Charlie Sands was climbing into his hunting blind. Typical of other blinds in use in the area, this one was a square, plywood, roofed box with windows on all sides, erected on a steel frame about 15 feet high. A hunter can sit inside and from this high elevation watch the brush country all around for a buck. With rifle slung over one shoulder and his pockets stuffed with lunch, ammo, and a thermos of coffee, Sands reached the top of the ladder when the blind door burst open and he was hit in the face with ball of snarling fur.

The next thing he knew, Sands was back on the ground with a broken arm and a busted rifle stock. The bobcat, which had been sleeping in the blind and had been surprised, was nowhere in sight.

"That cat was as big as a police dog," he assured me later. But in Texas the full-grown grayish-brown to reddish-brown cats measure about 3 feet, 6 inches from nose to tip of stub or bobbed tail, and weigh from 15 to 20 pounds. Occasionally somebody catches an old male in the chicken coop that will tip honest scales to perhaps 30 pounds. Although a rare northern bobcat has approached 60 pounds, such a giant in Texas would indeed be a phenomenon.

Nationwide, females average from only about 12 pounds in the Southeast to twice that in the northern states. The all-time record female may have been a 41-pounder killed in northern Michigan in 1970. Males vary from 20 pounds in southern locales to 30 pounds in the northern Midwest.

In January 1962 after a two-mile chase with his dog near Noblesville, New York, Frank Webb treed a bobcat larger than any of the 35 he had captured before. He carried the carcass to Cornell University where a zoology professor weighed the tremendous beast at 46 pounds, 10 ounces—8 pounds of which was fat. This male animal remains a New York state record, but not quite a North American record.

Another male trapped on Drummond Island in northern Lake Huron weighed 47 pounds. Reliable records of 55-pounders from Ohio and New Hampshire are cited in *The Bobcat of North America,* published in 1958 by the Wildlife Management Institute and compiled by predator researcher Stanley P. Young, then of the U.S. Fish and Wildlife Service. Young reports a 56-pound tom taken in New Mexico and one of 59 pounds from Nevada. But the most astonishing was a male taken in Colorado in 1951 weighing 69 pounds, which is close to the weight of a small female cougar.

Generally, however, bobcats are fairly small, light

In warmer climates, bobcats need far less hunting territory per cat than those in colder climates. We photographed this bobcat in California.

animals that make up in stamina and tenacity what they lack in bulk. Hunting bobcats with hounds has long been a popular pastime in America because the quarry is so challenging. A bobcat can run for hours well ahead of hounds, staying on top of a snow crust in which heavier dogs would flounder belly deep. It is always a mistake for a hound, no matter how big and fierce, to "catch" a bobcat on the ground. No dog is a match for this spitfire in a one-on-one contest. Taking on a bobcat, one veteran cat-hunter assured me, is like grabbing an electric fan with razor blades spinning at full speed.

**MATING, BIRTH, AND DEVELOPMENT.** People disagree on the ability of bobcat populations to survive intense trapping in northern America, though there are sure signs that bobcats cannot cope as well as coyotes. We now know that coyotes become still more wary and produce larger litters to make up for heavy losses in areas where they are heavily pressured. But the small size of the average bobcat litter, from two (and usually three) to five kittens, does not increase and seems not large enough to guarantee the species' survival through tough times. Virtually every serious study of bobcat reproduction has revealed a low survival rate of kittens.

Mating—amid screeching and howling heard over a great distance on the cold night air—takes place from late January in the deep South until March along the Canadian border. About 70 days later blind, helpless kittens are born in a secluded, usually remote hideaway. A bobcat's den may be under a mossy old deadfall, in a tangle of vines, or in piles of slashings left behind by timber cutters. But probably the majority of dens, and the most secure, are in old, standing hollow trees. Foresters tend to call these "wolf" trees; fortunately, they have no commercial value. The entrance to the hollow tree den may be fairly high above the ground, but is more likely to be in a cavity under the spreading root system.

A mother suckles her kittens for about three months. She must often leave them alone in the den for extended hunting trips, particularly during lean years when prey is less abundant than usual. Unlike a female wolf or coyote, a mother bobcat receives no help from a mate or from other relatives and must rear each litter alone. That probably explains the small litter size, as well as the low survival rate. It takes much time, energy, and skill to stalk and kill enough food for one adult and even more to supply a family of rapidly growing kittens.

Dependence on the female does not end when the

This young bobcat keeps a robin away from sibling kittens in this tree. The
bird was brought to the den by the mother and then claimed by this kitten.
This selfish trait is a key to survival.

young are able to leave the den. Instead they follow her for up to 10 months. Increasingly the group must roam farther and wider over the mother's home range so that they do not deplete the populations of mice and chipmunks, ground squirrels and grouse, hares and rabbits, possums and young raccoons, which they need to grow and accumulate fat for the winter ahead.

It is debatable whether a bobcat mother actually teaches kittens to hunt or whether hunting is a natural process fueled by gradual weaning and growing hunger. Kittens grow up at a time when prey species also are producing abundant young, so the hunting is easier than at other seasons.

Like all other wild carnivores, bobcat kittens spend long periods at play and in rough playfighting. This is a toughening as well as a learning activity, one of maneuvering, surprise, and pouncing, and of stalking—a skill vital to life in the future.

As soon as the young are agile enough, usually at age five or six weeks, the mother begins to bring live mice and active young rabbits to her brood for more serious training. She carries these in her soft mouth the way a Labrador retriever delivers a crippled duck. The kittens instinctively block the escape of these creatures, pouncing on them and seeming to toy with them, perhaps for a very long time, until the captive dies from shock or injuries. Even then the kittens may continue to worry the carcass and react to it almost as a plaything though they do not attempt to eat it. Keep in mind this is not a case of cruelty or sadism, but rather the introduction to predatory behavior that trains the species (and all other cats that behave similarly) to kill and survive.

Strangely, bobcats seem to be slow learners when it comes to killing prey. Playing with numerous captive creatures may go on for weeks before enough excitement is generated to inflict a killing bite. But as summer blends into autumn and small prey becomes increasingly more difficult to find, a young bobcat learns to kill things promptly or risk losing a meal. If more than one kitten has survived, there is also the matter of sibling rivalry or competition. The most adept, quickest killer in the litter wins the most food. The animal grows fastest to become even more superior and eventually a more successful hunter.

Only the most capable bobcat mother is able to raise more than a single kitten to the maturity needed to survive winter on its own. Among the pitfalls en route to adulthood are viral diseases, infections, parasites, and highway accidents. Areas near civilization also pose risk of predation by everything from dogs to great horned owls, as well as starvation and traps. Once full-grown, a bobcat's natural enemies are reduced to cougars and wolves.

Here a Texas bobcat has attacked and killed a jackrabbit its own size, demonstrating that it is a formidable predator. (Murry Burnham photo)

**FEEDING HABITS.** A bobcat's own life is one of feast or famine as long as it lives. Famine is most likely to occur during the first winter alone, then again late in life as the animal's physical reflexes slow down. Lagomorphs—rabbits and hares—are this spotted cat's staple. But a bobcat will eat anything it can catch. Its prey also includes frogs and lizards, snakes (even rattlesnakes), prairie dogs, domestic sheep and poultry and, unfortunately for its reputation in some places, venison.

There is no question that healthy, adult bobcats can and will kill fully grown deer as well as fawns. They will do so swiftly and efficiently. Since pioneer times, it has been widely believed that we would have more deer if we had fewer bobcats. Yet the facts do not support the bobcat's reputation as a deer-killer.

Texas A&M University researchers studying bobcats in Kleberg County, south Texas, found that when populations of cottontail rabbits and cotton rats were high, the cats fed on them to the exclusion of almost all other prey. When numbers of rats and rabbits declined one year, the predators were found to prey on 21 different species of small creatures, still ignoring the whitetail deer. A similar study completed in Arkansas at about the same time revealed similar findings, except that the bobcats there relied more heavily on mice.

A summer study of 32 bobcat stomachs at the Cu-

Next page: Particularly during the winter, bobcats will feed on carrion. This opportunist stands over the carcass of a deer pulled down by wolves.

Although capable of killing deer after a silent stalk and sudden pounce, bobcats tend to prey mainly on smaller animals.

sino Wildlife Experiment Station in northern Michigan revealed that 50 percent of the contents were rabbits or squirrels and only 20 percent meat from whitetail deer. During a second study of 53 bobcat stomachs carried out in winter, rabbit contents dropped to 13 percent while deer contents increased to 76 percent. Yet the same research showed that the bulk of that venison came from deer already dead or from those mortally wounded by hunters during the fall open season. A single frozen whitetail carcass can serve a bobcat for weeks, and the cat will eat to the final strip of flesh unless driven away.

During 15 years of following bobcat trails in winter, Michigan researcher El Harger logged 500 miles afoot. During that time he found only four instances of deer predation. One bobcat victim was a 120-pound buck; the other three were fawns less than a year old.

## THE BOBCAT'S MARK.
According to Harger, a bobcat kill is always easy to identify. A coyote or dog running wild pursues a deer, worrying it for a distance before pulling it down. But a bobcat pounces from hiding or at the end of a silent stalk, leaping on the head and shoulders. There it clings with the claws of all four feet, ripping the deer's throat open or breaking the neck by twisting the head around and back. Either way, the deer goes down within a few yards. But through the years, more than one woodcutter working around winter deer yards in northern forests has come across a bobcat killed by the sharp hoofs of a whitetail.

Like other wild cats, bobcats wander widely and hunt at night, but to what extent the creature is nocturnal is difficult to say. They do appear to be more nocturnal than cougars, but in daytime or darkness, the species has remarkably acute vision. The eye pupil, which narrows to a thin slit in bright light, gradually widens as light fades until it almost fills the iris by dark. Some scientists who have studied animal eyes believe that the bobcat has better depth-of-field vision than other predators such as wild dogs. These spotted cats can perceive prey in fine detail because their eyes have an abundant supply of both rod and cone cells. The rods are sensitive in poor or weak light and greatly increase night vision. Cones are used in good light.

## BOBCATS AND MAN.
Like some other predators, bobcats have a knack for often getting into the news. Attacks on people occur just often enough to invite wide attention. Probably none of those attacks is by animals seeking prey, but instead is by a rabid bobcat. Rather than adults, the targets are usually children playing outside in rural areas. Bobcats are among the carnivores most often found to have rabies, although the number of infected animals is low.

The late Ben East, well-known Michigan photographer, writer, and lecturer, related an interesting bobcat story to me. During the 1950s two of his friends, a husband and wife, were making a fishing float trip down the Au Sable River. With several trout in the canoe, the couple decided to pause on the bank, build a fire, and cook a shore dinner. As they paddled the craft beneath an overhanging tree limb on the bank, a bobcat dropped off the branch and landed in the bow. The wife bailed out over the stern at the same time the cat grabbed a trout and disappeared into the forest.

## AN ADAPTABLE SPECIES.
Bobcats are faring moderately well, although not equally well over their entire range in the face of proliferating civilization. At least according to one three-year study in Missouri in which 50 individuals were trapped and radio-collared, the species seems able to cope with modern forestry practices, even to any clear-cutting where the land quickly regenerates. Abandoned farms and fields also seem attractive to bobcats because of the abundance of small prey on which a cat can thrive.

No matter where it lives, *Lynx rufus* is one tough kitty that makes the American wilderness a more exciting place to explore.

The lynx of North America is a mysterious cat ▷ that roams evergreen forests of the north. Woodsmen find its tracks but rarely see the animal itself.

# CHAPTER THREE
# THE LYNX

Long-shanked and extremely quick and agile, a lynx hunts silently in the
densest cover. It is mostly a species of Alaska and Canada.

Early in the 1950s, I spent part of the summer fishing on Gods River in northeastern Manitoba for the trophy eastern brook trout that were large and plentiful then. As we were drifting down the clear, cold current, I was casting a streamer fly from a freighter canoe. We rounded a bend, and my Cree Indian guide Edgar Redhead touched my shoulder. "Link," he hissed, pointing to the left bank. There on an open gravel bar a lynx sat grooming itself. Because the rush of the river muffled any warning of our approach, we came quite near to the cat before it saw us.

Catching any wild predator by surprise is rare. And it is even more rare for a cat to be caught unaware. This one stood up on long legs, humped its back to stretch almost leisurely, and then catapulted out of sight in the way animals often do in Disney cartoons.

That evening Redhead and I huddled around a campfire and swapped wildlife stories, a good many of them about *Lynx canadensis,* the cat we had surprised. The guide admitted he had trapped far more than he had ever seen alive, concluding that the lynx is elusive and "as silent as smoke." I've never forgotten that phrase because it fits the subject so well. So, too, does the following passage from an old Cree Indian legend: "So the Great Chief who was master of all animals, gave Upweekis the Lynx a soft gray coat that is almost invisible in the woods, summer or winter, and made his feet large, and padded them with soft fur so that he is like the shadows that play, for you can neither see nor hear him."

**RANGE AND NUMBER.** We have no reliable index of the number of lynx across the United States in pioneer times. The lynx undoubtedly retreated from much of its original American range as logging, fires, and settlers cleared the dark forests from Maine westward to the Dakotas. The rapid elimination of prime lynx habitat, along with the basic intolerance of pioneering man, simply removed the lynx.

Unfortunately lynx are easily caught in traps, and from time to time fur prices have been high enough to stimulate plenty of trapping. Early French trappers used the term *loup cervier*—meaning "wolf attacks deer"—to describe both lynx and bobcats, because of the animals' similarity and because their ranges sometimes overlapped. By the late 18th century the lynx had already become scarce in western Europe, and its fur was in considerable demand in France.

Today the species occurs over most of upper North America. Its range south of Canada, however, which was never very deep into the continental United States, has been drastically reduced. The lynx is the only wild cat native to Yukon and Alaska. It wanders throughout the 49th state except on the Aleutian Islands, the islands of the Bering Sea, and some islands of Prince William Sound and southeastern Alaska. A similar lynx ranges across northeastern Europe and all across northern Siberia.

**PRINCIPAL PREY.** To date, we are not very well informed about the life history and behavior of the lynx. But we do know that the species is enormously dependent in particular on the snowshoe hare for its survival. Perhaps more so than any other American carnivore, the lynx's welfare thus rests on the availablity of a single prey species. An exception may be the black-footed ferret's reliance on the prairie dog.

Almost everywhere in the lynx's range, the cat's prosperity and survival depend on the snowshoe hare, which is cyclic. In years of abundant hares, lynx also are usually plentiful.

Lynx will eat ground squirrels and ptarmigan, lemmings and spruce grouse, but the snowshoe hare is the year-round mainstay.

As it happens, the snowshoe hare is characterized by cyclical highs and lows in population. The exact causes of these cycles remain unknown, but the mechanics are simple. During a period of, say, eight to ten years in a certain area, the population of hares increases to a peak. The creatures are everywhere. Suddenly there is a steep decline much more rapid than the buildup, and the hares seem to disappear. Then the whole process is repeated.

In some parts of Canada, snowshoe-hare cycles occur at such regular intervals that they are predictable. Elsewhere they are irregular with no recognizable pattern. Either way, lynx populations are cyclic right along with those of the snowshoe hares. Trapping records from the Hudson Bay Company from the 19th century reflect a distinct pattern of boom and bust in lynx pelts, so apparently the same factors were at work even then.

As hares multiply, lynx find more and more food available, and their own physical condition and chances of survival improve. Longevity is extended. More lynx kittens survive and fewer old animals perish of starvation or disease. Lynx numbers increase, often dramatically, right behind the increase in snowshoe hares. Likewise, the lynx population crashes soon after the inevitable decline in snowshoe hares as an adjustment to a drastically reduced food supply. Biologists like to use this hare-lynx relationship as a perfect example of a prey species regulating the numbers of a predator population, rather than vice versa.

In the Yukon, as many as nine of every ten lynx have perished after a snowshoe-hare collapse. While lynx are not normally long-distance travelers when food is plentiful, some survivors travel great distances—one in the Yukon as far as 450 miles—in a desperate search for food. Lynx are often seen in Arctic tundra areas at high elevation where they do not normally roam. That may explain why in 1965 Jack Lentfer, an Alaska Fish and Game Department biologist studying polar bears, was astonished to find a lynx feeding on a caribou at the edge of Arctic ice east of Point Barrow.

During a low cycle, a tagged Newfoundland animal covered 64 miles before being found near starvation. Lynx have been reported entering outposts and towns from Labrador to Fairbanks, Alaska, living (at least until apprehended) on local cats and dogs. Some of the lynx that appear periodically in Maine and in the northern Great Lakes states may be either rare natives, or they may be Canadian animals prowling far from home for something to eat.

During good times or bad, lynx populations are hard to estimate accurately. Tracking over winter snows can give some clues; but in summer the animals, which are as silent as morning mist, dissolve into their environments. Unlike ungulates, bears, or even bobcats, the lynx leaves behind few indelible tracks or markings. So, just possibly, they are not as scarce as their retiring behavior leads us to believe.

**PHYSICAL CHARACTERISTICS.** Somebody once called this northern feline an Arctic sphinx, which is appropriate enough. To me the lynx is the most ghostlike of all American wild cats. It is a long-shanked, short-tailed cat with pointed, tufted ears and a ruff of hair bearding its cheeks.

Superficially, *lynx canadensis* resembles its more southerly relative the bobcat. It weighs about the same and moves with the same fluid, effortless grace. Females generally weigh from 12 to 24 pounds, and males from 20 to 30 pounds. But there the similarity ends. A lynx's legs are noticeably longer than a bobcat's, and the feet are enormous by comparison. Especially in winter, each foot is further enlarged by a vigorous growth of stiff expanding hairs that function as broad snowshoes. With such oversize pads, even a large lynx can glide swiftly over fresh snow without floundering.

The ear tufts of a lynx are almost twice as long as those of a bobcat, giving the lynx what some people regard as a diabolical, or satanic, visage.

That is advantageous both for catching prey as well as for outrunning a wolf pack.

The tip of a lynx tail is always encircled with black. Ear tufts as long as two inches are twice as long as those of a bobcat (which may have none at all), giving what some regard as a diabolical or satanic look. Indians thought the tufts resembled feathers in a war bonnet. Also unique is the ruff or bib of long hair around the face and neck. Lynx tend to be gray or smokey in overall color, in contrast to the mottled black and reddish colors of the bobcat. The lynx is, to me, the more attractive and certainly the more mysterious of the two.

**BIRTH AND DEVELOPMENT.** In Alaska mating occurs in April, which may also be true across Canada. The female bears and rears the offspring on her own. Kittens are born deep beneath a windfall, far back in a rocky crevice or cave, or in a dense evergreen tangle. One litter of kittens was discovered in the Yukon in what was believed to be a black bear's den, vacated only weeks before by the bruins. Size of the litter is apparently determined by the mother's age as well as the current abundance of snowshoe hares. There may be a single kitten or any number up to ten. A female's first litter is usually smaller than later ones, but that same female may bear none at all if the period coincides with the bottom of the snowshoe-hare cycle. Charlie Abou, an Indian guide of the British Columbia Cassiars, told me of finding a litter of small kittens late in July, which may indicate that a lynx can have two litters during a boom period.

Lynx kittens are blind at birth, and those I have seen are not distinguishable from bobcats or some domestic kittens. In fact, I know of one lynx kitten that was taken from a wild den and raised to well past weaning with a brood of housecats. It eventually had to be separated from the others, not because it grew faster but because the others ganged up on it. Lynx kittens are buff-colored and marked with spots and stripes like those of an ocelot. Their eyes open within about two weeks of birth, and the spots and stripes disappear within three months.

How long lynx kittens remain with the mother is unclear, though probably it is well into the winter or until the kittens have attained their full skeletal growth and permanent teeth. Both are necessary if young lynx are to succeed as the solitary hunters they are. The kittens must accompany the females long enough to develop hunting skills and become familiar with their cold, northern wilderness environment. Several observers have noted that small kittens, when separated from the parent, have a cry quite similar to that of a gyrfalcon. During the crest of the snowshoe-

A lynx is an exciting animal to see, even if only in a fleeting blurred glimpse.

hare cycle, a female kitten might have her first litter as early as her first birthday.

During the mating season the lynx becomes highly vocal. I have never heard them, but fishing guide Edgar Redhead assured me that the call of a female in season was like something from a nightmare, similar to the blood-curdling cry of a human in agony. Despite their savage sound, lynx have never been known to attack a human under any circumstances. And because they are animals of deep forests rather than of clearings and the haunts of man, they do not prey on farmyard creatures. Still, these valuable furbearers were for too long the victims of a bounty system in many of the northern United States. The system lasted until about the time of World War II. Wisconsin, an otherwise progressive state regarding conservation, paid a $5 bounty on each animal killed until 1955.

**ENCOUNTERS IN THE WILD.** I have not had nearly enough opportunities to watch lynx in the wild. But my best chance ever came on my first trip to Denali (then Mount McKinley) National Park, Alaska, in the mid-1950s. At that time the park received only about 2,000 visitors during an entire summer, as compared to about 300,000 today, and overcrowding was never a problem. I made camp in or near what is today the Riley Creek Campground, an area that held a large Arctic ground-squirrel population.

At that time, campers routinely fed the ground squirrels that came begging around the few scattered tents, mine included. I cannot recall any regulations then against feeding the wildlife, with the exception

Pages 44–45: Many regard the lynx as the most furtive and retiring of North American cats.

Time out from the hunt allows a lazy yawn.

of bears. One damp and drizzly morning I sat inside my tent cleaning my camera, disconsolate over the prolonged rainy spell. One inquisitive ground squirrel came right up to the tent flap followed by another, when seemingly from nowhere a lynx pounced on one of them. It walked about 50 feet away and ate the squirrel. I was so startled, and so inexperienced then, that I forgot to use the camera in my hands.

During the next few days, which remained gloomy, the only bright spot was that same local lynx that continued to loiter in the campground vicinity. I believe I saw it at least once every day. It moved among the tents apparently without fear, even with people sitting in sight or talking inside the tents. But if anyone ever stepped outside or made a sudden movement inside, the cat would vanish and not return that day. The cat probably lived very well on the oversupply of squirrels while the bonanza lasted.

I recently spoke about the campground lynx with photographer Charlie Ott, who was a regular employee in Denali National Park for 25 years before retiring. Although a good many lynx live within the park, he never knew another one to become addicted to campground life. In a park where much of the wildlife is now accustomed to humans and is comparatively easy to see, Charlie regards the lynx as the most furtive, most retiring of all the native mammals.

In 1930 Charles Sheldon had unusual opportunites to see this secretive cat hunting when he spent an entire year in McKinley (Denali). Sheldon is the man most responsible for having this matchless portion of the Alaska Range designated a national sanctuary. After watching a lynx unsuccessfully stalk a band of a dozen Dall sheep rams on a steep ridge, he came upon that or possibly another cat that had caught a younger ram weighing about 120 pounds. From tracks in the snow, Sheldon could tell that the lynx dropped on the ram's back from a rock ledge as the ram stood in a ravine below. The lynx weighed only 20 pounds and seemed to be a starving animal. Sheldon either saw or knew of at least two other occasions when lynx tried to capture Dall sheep.

**A HUNTER.** The lynx is one carnivore that absolutely must be able to hunt well in brilliant daylight as well as in darkness. For a brief period each summer in Alaska and across northern Canada the sun never sets, while for a similar period during December there is very little daylight. The endless light of summer coincides with the time when all female lynx are busy feeding litters of kittens as well as themselves. Total darkness comes when populations of all prey species are falling to the year's low point.

During a recent summer, my wife Peggy and I were photographing grizzly bears in Denali National Park when we met a young couple who had just returned to park headquarters after a harrowing experience while backpacking. It was the last week in June, near the longest day of the year. Conscious of all the park warnings about bears, they had been careful to set up an overnight camp in a seemingly secure area where no bears were reported and where no bear sign was evident. They cached their food a safe, long distance away from the tent. Exhausted, the couple crawled into a sleeping bag and quickly fell asleep.

Suddenly one sat up wide awake and alarmed the other. Something outside was rubbing against the fabric of the mountain tent. "Bear!" they both thought, and were filled with terror.

The rubbing soon stopped and the culprit must have walked away. But some time passed before either was brave enough to unzip the flap and look outside. When they did, all they could see was a solitary lynx siitting on a rise about 100 feet away grooming itself. The hour was near midnight, but as the lynx posed there silhouetted against the deep salmon sky, there was no mistaking its identity. Nevertheless, the two had been too thoroughly frightened to fall asleep again that night. We were unable to convince them of their great luck in seeing that elusive cat at such close range.

The jaguar is the largest wild cat native to the New World. Once found north of the Mexican ▷ border, the species is now confined to isolated forest habitat in Central and South America.

46

# CHAPTER FOUR
# THE JAGUAR

In 1985, a terse, back-page news item from Mexico caught the attention of wildlife officials and others interested in nature. A ranch hand in Coahuila state claimed to have surprised a jaguar that was feeding on a dead cow and calf. The cat vanished immediately as soon as it saw the man. Back at the ranch, the cowpoke collected a posse of compadres with hounds and, they claimed, trailed the jaguar to the Rio Grande River. Presumably the cat swam the river and escaped into a remote section of Dimmit County, Texas.

Of course this could be a true account, but the odds against it are powerful. And the "jaguar" could have been a cougar.

For a long while after the New World was discovered by Columbus, the jaguar, *Panthera onca,* did indeed inhabit a fairly vast territory across the southern and southwestern United States in addition to northern Mexico. But no longer. For the jaguar's is a sad case of a magnificent native American animal that was wiped out before we realized it even existed.

Although it still survives in greatly reduced numbers and circumstances in Central and South America, the jaguar remains a cat nobody knows, except that it makes a striking fur coat for rich women and may be observed behind bars in zoos.

In 1922 Aldo Leopold made an expedition through the then wild delta of the Colorado River. A forester and naturalist who conceived the idea of saving huge wilderness areas in America, Leopold wanted especially to see one of the few jaguars that still clung to existence there at the time. Leopold later reported: "We saw neither hide nor hair of him, but his personality pervaded the wilderness; no living beast forgot his potential presence, for the price of unwariness was death."

During the next half century Aldo Leopold's son, zoologist A. Starker Leopold, made many trips into Mexico researching his fine book *Wildlife of Mexico* without seeing a wild jaguar alive. In 1959 he wrote: "Around the campfires of Mexico there is no animal more talked about, more romanticized and glamorized than *el tigre.* The chesty roar of a jaguar in the night

Jaguars rarely emerge from the shadows of green forests, which the sun does not easily penetrate. Unfortunately those forests are being destroyed, and the splendid cats are disappearing.

causes men to edge toward the blaze and draw *sarapes* tighter. It silences the yapping dogs and starts the tethered horses to milling. In announcing its mere presence in the blackness of the night, the jaguar puts the animal world on edge. For this reason it is the most exciting and interesting of all wild animals in Mexico."

The younger Leopold might have written instead, "of all the wild animals of North America."

Wandering jaguars probably generated a similarly tense atmosphere around the campfires of early woodsmen from Louisiana to Arizona before that land was settled. As late as 1845 John Woodhouse Audubon—son of John James, the great American bird painter—reported that jaguars still preyed upon horses and deer in the vicinity of San Antonio, Texas. Probably using a captive jaguar there as a reference, he completed the first painting of a jaguar by a white artist. It hangs today in the Amon Carter Museum in Fort Worth.

The last jaguar actually seen in Texas was dispatched in 1946, details unknown, while the last individual known in the United States was shot three years later in Arizona. Perhaps not as strange as it may sound, some local Mexicans believe that the 1985 jaguar incident on the Rio Grande was only a ruse to cover up a drug-smuggling operation nearby. How life and times have changed.

**THE PHYSICAL SPECIMEN.** Among the few facts we know about jaguars is that they are the mightiest of the New World wild cats. At a glance the jaguar closely resembles the leopard of Asia and Africa, though on closer inspection there are distinct differences. The jaguar has a broader, more massive head and a shorter tail. The background color of a jaguar's fur is light cinnamon to golden buff. Its coat is covered with numerous black blotches, many of which are grouped on the flanks to form rosettes that contain one or more central spots. These spots do not appear on the leopard.

Within the jaguar's original wide range from Sonora, Mexico, southward to central Patagonia, Argentina, several similar subspecies have evolved. Extremely rare all-black or dark brown jaguars—on which the rosettes are nevertheless faintly visible—exist in the wettest, forested parts of their range.

Adult jaguars measure 27 to 30 inches tall at the

shoulder and weigh between 125 and 250 pounds. Males average larger than females. Perhaps the cats of central South America are generally heavier than those north of Panama, but too few have been accurately weighed by scientists in various locales to make firm comparisons. There seems to be no definite, annual mating season. Nor are there definite figures on wild litters, though captive females have produced litters of two to four cubs, which are spotted at birth.

Jaguars are capable tree climbers. I once saw a jaguar bound in a flash from the base of a tree 30 feet up into its crown, making hardly a sound. But the species spends most of its time on the ground stalking peccary herds and deer, waiting to pick off any stragglers. Fish, turtles, and snakes are easily hooked out of the water with swift retractable claws used as gaffs. Jaguars have also been found roaming ocean beaches on hot spring nights to excavate nests of sea-turtle eggs. They eat both eggs and hatchlings.

**RANGE AND HABITAT.** The consensus has been that jaguars live on small home ranges of roughly 15 to 20 square miles, although such figures tend to be only educated guesses and are not without exception. In 1955, for example, a solitary male was killed near the southern end of the San Pedro Martir mountain

range in Baja, California. To reach that remote place, the animal must have crossed the entire Sonoran Desert as well as the Colorado River before trekking south still another 100 miles. Altogether it was a trip of about 500 miles from the nearest regularly occupied jaguar range.

*El tigre* thrives in such diverse habitats as rain and riverine forests, highlands, and especially lowlands, as well as in dry brush and tall grass savannas with sufficient cover. Access to open water in which it can swim, cool off, and pursue prey must be part of the territory. Underdeveloped coastal areas are also usually good habitat. Jaguars take to water more willingly than any other New World cat.

**JAGUARS OF COCKSCOMB BASIN.** In March 1983 a trim, energetic wildlife biologist arrived in Belize, Central America. In the years since his arrival Alan Rabinowitz has needed every bit of his energy and dedication because Belize is not your ordinary kind of place and studying its jaguars—his mission—would have been a formidable undertaking under the best of circumstances. The biologist's experiences there sound almost like a strange and sometimes hair-raising adventure. But they are, nonetheless, factual.

Belize is a small country about the size of Massa-

Excellent tree climbers despite their large size,
jaguars spend most of their time hunting on the
ground for smaller prey.

chusetts, just south of Yucatan, Mexico. It is hemmed against the Caribbean by Guatemala on the west and Honduras to the south. Formerly it was British Honduras, regarded as a dank, unprofitable backwater by the British landlords. Her Majesty's government granted this last unprofitable colony its independence in 1981.

Only about 150,000 people live in Belize, most of African descent, and are crowded into communities along the hot mosquito coast. This extremely rugged country is heavily forested, and has a most unpleasant climate. I have made two trips to Belize and have been thoroughly dry and comfortable there on only fleeting occasions. Belize residents, both human and wild, endure an eight- to nine-month rainy season. The capital, Belize City, has been ravaged by hurricanes twice in recent decades.

So Rabinowitz did not select a natural paradise in which to work. But he had one vital thing in his favor. The same conditions that made human existence unpleasant also made the country the last haven for many kinds of wildlife. For example, Baird's tapir and the equally rare black howler monkey are making a last stand there. And the environment is still able to support a fair population of jaguars, particularly in dense jungle. The cats, however, are not beloved of the Mayan Indians, whose cattle jaguars sometimes kill. Ironically, the Maya of another time built great stone temples and pyramids to worship the jaguar.

Rabinowitz selected a section of Belize known as the Cockscomb Basin, a remote tumbled forest looming over dense undergrowth where a riot of green plants and flame-colored birds compete to live. The Basin is also inhabited by scattered Mayan families. Rabinowitz's plan was to capture alive and radio-collar as many cats as possible to get some idea of population density, where and how far jaguars travel, and what they eat. He also wondered if all jaguars are cattle killers as most Belizeans believed, or if only a few individuals did most of the damage.

Radio-collaring and tracking animals has been a proven method of studying everything from muskrats to rhinos since the 1950s. But Rabinowitz's Indian neighbors thought that anyone tracking jaguars except to kill them and sell the hides was mad, possibly dangerous, and they avoided him. The first wooden traps he built were chewed apart by the first cats that entered them. Months passed before Rabinowitz actually caught his first jaguar alive, and for that he had to hire Bader Hassan, who had been a professional jaguar hunter until the sport was discontinued in Belize. Hassan, it happened, still owned a pack of hounds.

One steamy day, while running a jaguar with Hassan's dogs, Rabinowitz's native assistant was bitten by a fer-de-lance, a deadly poisonous viper that is only one of the hazards of life in the Belizean jungles. They treated the man with antivenin, flew him to a hospital in Belize City for care and recovery, and then returned to the chase. It ended with the capture of an adult male, which was collared and released. They called that first jaguar Ah Puch, which was the Mayan god of death. The name became doubly apt because the man bitten by the fer-de-lance promptly died when his wife removed him from the hospital and turned to a native shaman—or healer—for treatment.

Rabinowitz decided to accelerate his trapping by using the Hassan hounds in conjunction with stronger traps of iron framing, which were also somewhat portable. He succeeded in capturing four more jaguar males in the traps. Another male and a single female—the only female taken in the study and a known cattle killer—were caught using the dogs in northern Belize. These were released in Cockscomb Basin.

A mystic wandered into Rabinowitz's camp one day and claimed that for remuneration he could communicate with jaguars. After burning copal (an aromatic tropical resin) and speaking in a strange tongue, he assured the biologist that another jaguar would be caught within three days.

The man was at least half right. A cat was indeed reported entrapped, and Rabinowitz raced to the spot on his dirt bike followed by curious Indians. But it turned out to be Ah Puch. This time, both of its top canine teeth had been broken off as the animal tried to free itself from the trap.

The biologist tranquilized the animal and pried open its jaws to inspect the teeth, which he could see were badly infected as well as broken. The jaguar closed down suddenly on a finger and clawed his shoulder. Rabinowitz remembers talking crazily to the huge cat, perhaps crying, and then carrying it over his shoulder back to his cabin for treatment of the infection. But as the Indians watched in awe, the jaguar died in the biologist's arms. After that Rabinowitz was known as "el tigre hombre," the jaguar man.

The bad luck didn't end there. On two occasions a fer-de-lance struck at Rabinowitz but fortunately hit only loose clothing. One day Rabinowitz fell almost 60 feet from a tree but rose unhurt. Another time, while Rabinowitz was returning from a flight to track his radio-collared cats from the air, a thunderstorm pounded the airstrip as the plane circled to land. Trying to climb from the dangerous situation, the pilot clipped trees and lost the left wing. This flipped the light craft into a 150-foot nosedive directly toward the ground. Yet the thick forest canopy cushioned the impact and saved two lives.

The two men had to follow a jungle creek for several

hours to reach help. All the while Rabinowitz suffered from a broken nose and the increasing pain of several blood clots. While being stitched up that night in a Belize hospital, Rabinowitz recalls wondering if the Indians weren't right about his being mad.

Eventually some of the Indians became somewhat fascinated by the jaguar study, perhaps because they couldn't comprehend locating any live animals without actually seeing them. A few eventually understood the concept of the radio collars and even learned to operate the equipment. In fact the one female jaguar that was trapped wandered back to her old haunts and was killed there by waiting Indians who had learned the radio-tracking "secret."

Though capturing wild jaguars was the toughest and most time-consuming part of the study, tracking them after release was almost as difficult. In such dense habitat, signals—a different one for each collared cat—rarely carried more than a mile or two over level ground. If a jaguar happened to stop in a ravine or depression to rest, the signal distance was reduced to a half mile or less. In more open country as on the African plains for example, the same kind of collars would transmit signals for up to 10 miles.

But Alan Rabinowitz persisted, and by tracking his five collared adult males for as long as 14 months he managed to gather most of what meager scientific information on the species presently exists.

**TYPICAL JAGUAR PREY.** Rabinowitz learned much about jaguar feeding preferences in the Basin. He found that they will take whatever they can get and depend much more on small rather than large animals. Armadillos, agoutis, pacas, rodents, and iguanas are all typical jaguar prey. In tropical areas a jaguar's diet may be further enriched by monkeys, sloths, capybaras, tapirs, and caimans. But the smaller ani-

mals are more available and much easier to subdue without injury than stronger tapir and deer (both whitetail and brocket deer share the jaguar's range in Belize).

The Cockscomb Basin males had overlapping territories of from 11 to 16 square miles each—smaller home ranges than previously suspected. Keep in mind, though, that the *density* of cats in Cockscomb Basin may be greater than anywhere else. Such density might suggest frequent contact and probably fighting. Yet during a total of ten captures, in addition to occasional sightings of collared males, Rabinowitz found only one case of scarring or injury that might have been related to aggression. Females and a small number of cougars also ranged in the Cockscomb study area, but the extent of their travels is not clear.

As with many other cats and carnivores, jaguars seem to use vocalization and defecation to announce their presence and mark territorial claims. Jaguars avoid other cats, as well as the pugmarks and piles of feces left by other cats. Alan Rabinowitz found numerous piles of droppings along trails and old logging roads, conspicuous sites hard to miss for a wild animal with keen senses. He also discovered scrapes and claw marks both of adult male jaguars and of a cougar traveling in areas of overlap. Though unfathomable to humans, these, too, are probably signs obvious in meaning to solitary predators that need to avoid one another.

The importance of the behavior trait *avoidance* cannot be overestimated, especially in a place such as Belize. It is true that jaguars, the top of the food chain, have no enemies here except for human hunters. According to Rabinowitz, however, these tropical forests host a great array of parasitic and microbial fauna ready to take advantage of any injury or wound that provides an opening to the body. In the long run

The endangered Baird's tapir plays a minor role as a prey species of the jaguar.

Jaguars more often hunt the tree iguana. The cats are not the wanton killers claimed in folklore.

Normally solitary mammals, a male and female jaguar
may keep company for a while during the breeding matchup, which
can occur at any time of the year.

then, there may be no winner in a territorial or mating fight between two tom jaguars since any open wound could eventually be fatal. Remember how an infection in the jaw killed Ah Puch? Thus, it is possible that animals that avoid injury may well live the longest.

**THE JAGUAR HUNTERS.** Sport hunting has been blamed for killing too many jaguars and making them scarce. But the toll of the cats taken by sportsmen has always been small. The target is simply too elusive and inaccessible, and the hunting itself too gruelling and physically unpleasant. Best-known of the old-time jaguar hunters, due to his self-promoting lecture tours across the United States in the 1940s and 1950s, was one Sasha Siemel. He claimed to have killed more than 230 of the "ferocious predators" armed only with a spear or bow and arrow.

Most of the kills were made in the Mato Grosso, the great, featureless grasslands of central South America. Siemel liked to say that he was only hunting cattle-killing cats and man-eaters, and he used motion pictures to show how he did it. He claimed to track the jaguar with dogs until it finally came to bay. Then when the cornered cat charged, it would be impaled on the end of a spear. Siemel's method would take

plenty of nerve, yet one wonders about the curious psychological need to destroy 230 of the splendid animals.

In 1963 I met an Indian hunter in Cienaga de Zapatosa, Colombia, who also killed jaguars with only hounds and a spear. Macedonio Polo assured me that his body count was more than 50 jaguars in about 10 years of pursuing them barefoot and by dugout canoe. His reasons were purely commercial. He was able to support a large family very well on the proceeds from the sale of the spotted hides alone.

Around a campfire one night, Polo described how some Indians living deeper in the Colombia interior often captured jaguars. The first step was to trap a number of parrots or macaws or other large, noisy birds alive. These were staked out around a tree in which the hunter hid with a spear. Another parrot was tied by one foot and held in the tree with the hunter.

Days might pass before a cat discovered the staked-out parrots, which kept up a steady chattering and squawking. But these hunters had nothing but time and plenty of patience. Eventually a wandering jaguar (or ocelot) was lured to the spot and ate the birds. When it was nearly finished, the hunter dangled his captive bird downward by one foot until the cat moved

My first days (crouched in a tree blind) waiting for a jaguar were tedious and frustrating. The first animal to appear was a white-lipped peccary, which spotted me immediately and ran off. Then finally one morning this jaguar appeared and vanished an instant later.

directly beneath him. Then he let fly with his spear. It is difficult not to admire such ingenuity and raw courage. Polo also told me he knew of at least two Indian hunters who went jaguar hunting thus and were never seen again.

Nowadays, I hear, natives are hunting those same jungles in a different way. At night they creep in 4-wheel-drive vehicles along the network of roads made by timber cutters, spraying the surrounding cover slowly with the beam of a bright spotlight. When the beam illuminates a pair of ruby-red eyes out in the black, they shoot, then check later to see if it was a jaguar or only another hapless jungle creature.

Macedonio Polo told me some other interesting things about jaguars. The animals are often credited with extraordinary feats of strength such as dragging a full-grown ox into the crown of a tree or swimming across a jungle river with a horse grasped in its jaws. During an expedition to Brazil, Teddy Roosevelt was told that one cat had dragged a large horse more than a mile from where its neck was broken. But Polo said jaguars always eat prey very near to where they kill it unless followed by humans, in which case they run away.

Another popular belief is that jaguars drop down

out of trees onto the backs of their victims. During a lifetime spent in jaguar habitat, Polo never knew a cat to drop down from a tree ambush.

Polo had yet another revealing story to tell and chuckle about. A successful businessman from Bogota, Colombia's capital, bought wilderness land in the Cienaga de Zapatosa area and had it cleared for a ranching operation. He also offered Polo a bounty for each jaguar moved away from the vicinity. Unable to find any jaguar right there, Polo killed a female with a cub a long distance away, but nevertheless did not hesitate to collect the bounty for two head. When the land clearing was completed, a shipment of sleek, fat beef cattle arrived one day from Spain to stock the ranch.

According to Polo all the imported cattle were dead or had vanished within a month. Of course jaguars were blamed, but Polo assured me that cats did not get a single one. "Then what?" I asked. "Insects, fever, and hungry human neighbors," he replied, chuckling again.

I asked Polo about cubs and dens, and he replied that both were hard to find. Females were extremely secretive and must have located their natal dens in areas of the forest difficult for even the best Indian woodsmen to penetrate. Judging from the roaring and caterwauling sounds of mating he had heard, mostly at night, Polo said the mating season was continuous. We know from captive animals in zoos that the gestation period is about 100 days. Polo knew of no female in his part of Colombia to have more than two cubs. Young cats were much easier to catch and corner with packs of dogs, and these comprised most of the 50 or so he had killed.

During my long trip into the headwaters of the Magdalena River of Colombia in 1963, I heard for the first time the resonant vocalizations of the animals as well as those of a hunting guide imitating them. From a tree-stand well inside the edge of a gloomy jungle, I spent several nights with Victor Villareal. From time to time far out in the darkness I could hear a sound best described as a rasping grunt. It came as if from an animal walking, perhaps approaching. Villareal imitated the sound very well by grunting into

a wide-mouthed earthernware pot. The cat he called was never tempted quite close enough to be seen in the weak spotlight we carried, but tracks on the ground at daybreak came to within 100 feet of our tree—from behind!

Because of the considerable value of jaguar hides, jaguar-calling is a valuable skill in many remote corners of Latin America. Some backcountry Indians are able to use their mouths and vocal chords alone, others use jugs or bottles for resonance. In a village near Orange Walk, Belize, an old man showed me a primitive call he claimed had accounted for "many, many tigers." It was a dried gourd slit open and covered most likely with deer hide. He rubbed a band cut from an auto inner tube against the gourd. The sound was not loud, but may have worked. Alan Rabinowitz described a similar call made from a gourd and deerskin, with which a waxed banana skin was used. But in Cockscomb Basin, he did not hear jaguars calling with

This jaguar was driven into a Colombian swamp by a pack of ranch dogs. I photographed the animal from a dugout canoe.

57

enough regularity to understand their vocalization.

Whereas Macedonio Polo hunted for the valuable spotted hides, others have hunted jaguars for a living by taking out sportsmen who need a jaguar skull and skin to complete a big-game trophy collection. Capable jaguar guides and outfitters have always been in great demand, though it is now illegal to import spotted cat skins into the United States. Typical of these men was J. F. Ferreira of Sinaloa, Mexico, who accounted for about 60 jaguars during his career. Ferreira concluded that in his region male jaguars lived in very small territories, which they vigorously defended against other male intruders. He also considered the cats entirely nocturnal. His strategy was to take a station in a known male's territory and then roar in imitation of another *tigre*. The method worked well for him, especially if a female in estrus was also roaming the vicinity.

No doubt the most famous of the Mexican jaguar hunters were the Lee brothers of Tucson, Arizona, who owned a pack of cat hounds at least the equal of any ever assembled. The Lees hunted widely south of the border, in the Mexican states of Campeche, Tabasco, and northern Chiapas. But their most successful hunts were along the Pacific coast of Nayarit, where they often had to follow the dogs through mangrove swamps by boat. Both the jaguars and ocelots there were at home in and around the water, so pursuing the cats in that aquatic environment had few dull moments. Even during the best of hunting trips, the jaguar trophy never came easily.

**JAGUARS AND MAN.** Alan Rabinowitz blames poor ranching practices for at least some of the jaguar predation that has backcountry residents up in arms. These locals frequently allow growing herds of livestock to range too far and for too long unattended in jaguar country. Common cattle thievery by humans is also frequently blamed on the cats. According to Rabinowitz some ranchers shoot any jaguars in their area before cattle killing starts, but this policy could be counterproductive. Post-mortem examinations made of jaguars caught preying on cattle revealed that three of every four had been shot previously, usually with shotguns, and may thus have been rendered unable to kill wild game. Two "cattle killers" I once examined in Colombia were both carrying bits of metal fired from shotguns.

As a preventive of livestock killing, natives hunt jaguars. Ironically, three of four jaguars finally killed have been previously wounded by firearms, which may have rendered the big cats incapable of surviving on wild prey alone.

Although *Panthera onca* remains the least studied, least understood of all the world's large cats, the work of Alan Rabinowitz resulted in a particularly splendid accomplishment. In 1985 Belize's Cockscomb Basin was redesignated the Cockscomb Basin National Park, the only such sanctuary on earth specifically set aside for the jaguar.

In too many places elsewhere, timber cutting and slash-and-burn agriculture is reducing jaguar habitat to eroded wasteland. Most governments have been slow to realize that tropical forests are by far most valuable to their economy when left alone and intact. Soon after all the trees are cut for a quick profit, the jaguars and their prey will vanish forever. But that is only the beginning of the tragedy. The first torrential rains will wash away the unprotected, thin topsoil and its nutrients, leaving bare unproductive land on which nothing grows.

Peggy and I have spent a good bit of time wandering and photographing wildlife in some of the best jaguar country left in the Americas, from Nyarit to Iguazu Falls National Park, Brazil. Except for a few biologists and hunting guides such as Macedonio Polo and the Lee brothers, we have never found anyone who has actually seen a wild one alive. It is far easier to find people who truly believe the cats are man-eaters.

Undoubtedly native hunters have been killed by cornered or wounded jaguars, though we can find no truly authenticated cases of unprovoked attacks. Nor could Starker Leopold when researching his book *Wildlife of Mexico*, though he devoted much time to research. So in this respect *Panthera onca* and the other American cats differ from their Old World relatives— lions, tigers, and, leopards—in having never become confirmed man-eaters. Knowing that, I suppose, may take away some of the jaguar's considerable mystique.

Today there may be increasing interest in saving the jaguar, not only in Belize but elsewhere in its range. The World Wildlife Fund and the New York Zoological Society, among other conservation agencies, have allocated considerable funding to benefit the jaguar.

Is it really worth it? As Alan Rabinowitz recently noted, "The jaguar, caught up in a destiny over which it has no control, is a splendid creature with every right to survive on earth."

Among the most striking of all the world's cats, ▷
the ocelot is relentlessly hunted for its soft
luxurious fur.

# CHAPTER FIVE
# SMALL TROPICAL CATS

With adults weighing from 20 to 35 pounds, ocelots are heavier than the
margay and the jaguarundi. Markings on no two ocelot coats are ever
exactly alike.

As recently as a century ago all of the native North American wild cats, except for the Canadian lynx, still roamed free in Texas or elsewhere in the American Southwest. All but the lynx and perhaps the jaguar still do today. We have already covered the cougar, bobcat, lynx, and jaguar. That leaves the trio of tropical species: ocelot, margay, and jaguarundi. None of these is well known, and the status of each is uncertain at best. These cats are very seldom sighted in the wild anywhere, let alone north of the Mexican border. Indeed there is only one sure record for the margay; a single specimen was shot near Eagle Pass, Texas, by a Colonel S. Cooper in 1881.

**THE OCELOT.** The ocelot, *Leopardus pardalis,* is by far the most strikingly beautiful and largest of the three. It can be classified as a medium-size wild cat. Though no two ocelot pelts are exactly alike, the fur is so soft and luxurious that ocelots are among the most heavily hunted of all the world's wild cats. No wonder the ocelot has become rare over large portions of its original range, which extended from the Rio Grande Valley southward through coastal Mexico and Central America to most forestlands of middle and South America.

The World Wildlife Fund—an international conservation organization dedicated to saving endangered species and habitats—estimates that during the 1970s about 4,000 ocelot pelts were illegally exported from the Chaco region of Brazil alone. In the same period an additional 300 pelts were poached in Bolivia and shipped to furriers in Europe. That was a time when every show-business celebrity and countess on the continent simply had to have an ocelot coat to appear in public. Of course this fad was devastating for the cats.

Ocelots can live in dry bush country, even on the edge of deserts, and those that do have a grayish background color. But the majority dwell in damp mountain forests or humid jungles and show an ochre or orange background. The ocelot is seldom a creature of open, exposed country, except for marshlands. Expert furriers can look at an ocelot skin and usually tell exactly where it originated.

The head of the cat has numerous small dark spots and twin stripes on each cheek. Four or five black stripes run parallel along the back, while elongated spots run in rows obliquely down the sides of the body. Adult males may measure almost four feet from nose to tip of tail; females average about three feet. Adult weight ranges from 20 to 35 pounds.

We know very little about the habits and life history of ocelots, especially in the northern portion of their range. No scientist has yet made a detailed study of them. But most biologists and backcountry natives in ocelot country believe the animals are either mostly or entirely nocturnal, sleeping during daylight hours and hunting after dark. Biologist Edward A. Goldman, working in Panama and searching on horseback for the cats, found ocelots sleeping on tree limbs. When thus exposed, the cats usually tried to escape by climbing slowly upward until out of sight. When ocelots were suddenly surprised by a man on foot, however, they would drop to the ground and race away. Goldman never once encountered an ocelot when hunting with a lamp at night. Depending on the environment, ocelots may also spend daylight hours sleeping unseen in hollow trees, in caves, or even in dense prickly brush.

Naturalist Frank M. Chapman, studying on Barro Colorado Island in Gatun Lake, Panama, observed that ocelots may use the trails of humans to wander and hunt at night. As a result, Chapman took the first and among the best photographs of these spotted cats in the wild by setting up a flash camera triggered by a trip wire across a frequently used jungle path. The technique worked well. But on examining his exposed film, he noticed that almost all of the cats had been photographed with the leading foot about ten inches from the ground. Since that would have been an unusual posture on the clear, smooth surface of the path, Chapman concluded that the ocelots were well able to sense the thin wire obstacle in the dark and were attempting to clear it.

The ocelot's ability to hunt and survive in tall forests where neither sunlight nor moonlight can penetrate indicates the keenest, most well developed senses. Although ocelots are arboreal at times, they are not especially quick or skillful tree climbers. They hunt largely on the ground, not by ambushing but by running down their prey. They are not at all shy of hitting the water and can swim very well. I once saw an ocelot swimming the Parismina River in eastern coastal Costa Rica so rapidly that, even in our dugout

Sharing the ocelot's jungle habitat are the crested guan (top), the green tree snake (middle), and the red-lored parrot (bottom).

canoe, we could not approach near enough for photographs.

Probably the best account of this species—called *ocelote* or *tigrillo* (in Spanish) or *xacxicin* (Mayan)—is that of Swiss naturalist J. R. Rengger, who spent six years in Paraguay during the 1830s. Rengger thought ocelots never enterd water willingly, but were forced to siwm often by the great seasonal flooding along the Rio Paraguay. After watching the cats climb

and occasionally jump from tree to tree, Rengger wrote that they were not as agile off the ground as, for exaple, much larger cougars.

Rengger was also the first to report that ocelots live year-round in male/female pairs within well-defined territories, a unique arrangement among felines. The two do not roam about together although they may not be far separated. In Rengger's observations, no territory ever contained more than one couple, and male and female ocelot share responsibility for marking it by spraying with urine, by defecation, and by other means. The two cats communicate by mewing, which intensifies during the mating season to shrill yowling and, at night, caterwauling.

Actual breeding seasons vary greatly from place to place without pattern. Gestation is about 70 days. Litters of kittens have been found in the lower Rio Grande Valley of Texas in autumn, indicating that the breeding there occurred in summer. The nursery is usually a cave, hollow tree, or secluded depression. Two is the typical number in a litter, though there is at least one record in Mexico of four. Ocelot kittens are treated, weaned, and raised much in the manner of bobcat kittens. We have no data on survival rates.

Prey animals of the ocelot include mostly small mammals such as rodents, agoutis, pacas, and rabbits. Ocelots also can catch some birds such as guans and chachalacas, even when flushing or on the wing. In certain parts of their range ocelots eat small monkeys, and almost everywhere reptiles make up part of an ocelot's diet. One day in 1916 naturalist Edward W. Nelson was riding a horse along a forest trail in Chiapas, Mexico. Suddenly an ocelot jumped from a brushy ravine and climbed into a tree beside the trail. Nelson shot it with a revolver and, on retrieving the cat, found it had been feeding on a recently killed seven-foot boa constrictor. The animal had already consumed the head and neck, and Nelson's approach had interrupted its meal. A reptile of that size could not have been easy to kill.

During a trip I took to Belize in the 1970s, Hernando Brown—an old river guide who had both trapped and hunted ocelots with hounds for many years—told me he had often found kills of brocket deer and that he also used these animals as bait, around which he placed steel traps. Brown said some ocelots might live on larger birds, such as crested guans, currasows, and turkeys, or even gorge on the young herons and egrets that fall from nests in the large rookeries of Crooked Tree Lagoon. In nearby Yucatan a government forester told of seeing an ocelot catch an ocellated turkey on its roost and then fall with the struggling bird from the limb to the ground.

For the most part, ocelots shun civilization and do

Although ocelot kittens make appealing household pets, their unfortunate popularity has caused a drain in the wild kitten population that may be as significant as fur trapping. Adult ocelots usually prove to be much less desirable pets, however.

not survive long around human settlements, even remote ones. But just often enough an individual or a pair begin to break into hen houses or develop a taste for piglets that are easily stolen from flimsy sties. Neither act is very popular, and often as not, a lot of cats are killed in pursuit of the single culprit.

Unfortunately, the species has another disadvantage in addition to its attractive coat and occasional liking for domestic fowl. Particularly when captured very young, an ocelot is likely to have a much milder disposition than most other wild cats and to survive far better in captivity. Thus there is also a great market today for ocelots as house pets.

Keeping an ocelot would be cruel practice for all but a few dedicated humans who truly understand and can tolerate wild cat behavior. Capturing kittens alive drains wild populations as surely as trapping them. They usually fare poorly in inexperienced hands. If ocelot kittens survive to adulthood, they are soon disposed of because of increasingly unpredictable behavior. Their urine-spraying instincts can soon make a home or apartment unlivable. Too many are defanged, declawed, or both, with most winding up in an animal shelter or local zoo where there are already too many ocelots on hand.

The best thing we can do for *Leopardus pardalis* is to leave it alone in the wild and save all of the wild habitat that remains.

---

**THE MARGAY.** The margay, *Leopardus wiedii*—also called *tigrillo*, tiger cat, or *chulul*—is the ocelot's closest look-alike relative. It is smaller than an ocelot and about the same size or a little larger than the average domestic cat, but is much more lithe and agile than either. Its head is shorter and rounder than the ocelot's, and the dark eyes are noticeably larger. A margay's tail is longer, with longer hair, and not as pointed as its larger cousin's. The color and pattern of its fur is nearly identical to that of the ocelot, however, which sometimes makes positive identification difficult with only a brief glimpse.

An exclusive forest dweller, the margay may be the most accomplished acrobat of all wild cats, bar none.

Smaller than the ocelot, the margay is more lithe and agile, and more
comfortable above the ground. The margay is so agile in trees that central
American bush dwellers call it the *monkey cat.*

Its joints are anatomically adapted to allow the hind
feet to rotate inward to a full 180 degrees. Thus the
margay can hang from a branch by its hind feet or its
"wrists," or move along a branch while hanging upside
down in the manner of a sloth. It can even jump off
a tree limb, or release its hold in the sloth position,
and still catch itself with extended claws on another
branch some distance below. Hernando Brown told
me that on a number of occasions he at first mistook
a margay moving in a tree crown for an extremely
agile spider monkey.

Like no other cat, tigrillo can run headfirst down
a tree trunk at a gallop or with a diagonal walk. Beyond its astounding agility, though, next to nothing
is understood about the margay's life in the wild.
Dr. Inga Poglayen, formerly wildlife curator of the
Arizona-Sonora Desert Museum in Tucson, sums up
what we know: "It is likely that this strangely different
cat spends nearly all of its time in the tree canopy,
there hunting its prey of birds, squirrels, marsupial
rats, and small monkeys." The animals are especially
adapted to a forest life, hunting all kinds of bird
nests and feeding on the nestlings.

Another reason for our lack of knowledge is that

the margay is an exceedingly rare mammal, very elusive, or possibly both. In the early part of the 20th
century, naturalists Edward Nelson and Edward
Goldman collected 15,000 native mammals from
every part of Mexico in a thorough survey of that
country's wildlife resource. Of that number, only two
margays were taken. Scientist A. Starker Leopold
notes in his book, *Wildlilfe of Mexico,* that he knew
of only two additional specimens taken in that country, although the margay's range is probably similar
to the ocelot's.

In South America, this species ranges from Panama
and northern Colombia southward to Peru, Paraguay,
and northern Argentina. But nowhere is it plentiful.
J. R. Rengger, who had considerable exposure to ocelots, could obtain only one margay specimen in Paraguay. Another naturalist, G. F. Gaumer, reported
that a margay brought to him by Indians as a kitten
in 1917 became a clever and somewhat affectionate
pet. At least, according to Gaumer, "it waged incessant war against the mice and rats that infested the
house."

Sad as it may seem, that pretty well covers all we
know of this most mysterious and fascinating cat.

A plain-colored, mostly ground-dwelling wild cat, the jaguarundi may still roam in the lower Rio Grande valley. At first glance, it somewhat resembles a mink or weasel.

**THE JAGUARUNDI.** Early on a humid April morning in 1986, Peggy and I began a steep hike to Xunantunich, once a busy city of the Mayan Indians in the Maya Mountains of western Belize. We had crossed the Mopan River by ferry, shouldered day-packs, and followed a trail through scrub jungle that would eventually lead to the anicent complex of crumbling pyramids and palaces, homes, and cere-monial plazas. Xunantunich is not nearly as well-known as Tikal in Guatemala, or Chichen Itza, Tul-um, and Palenque in Mexico; but its remote location makes it just as haunting and just as exciting to visit. We hiked slowly in the heat, pausing to watch a myr-iad of bright butterflies and identify the astonishing variety of jungle birds. Halfway from river to ruins, a strange dark animal that resembled a mink bounded across the trail before us. Puzzled for a moment, I focused binoculars on the creature and immediately realized I was watching a jaguarundi.

In all the world no cat is more un-catlike in ap-pearance than the jaguarundi, *Herpailurus yagouar-oundi*, also called the *eyra* or (in Mexico) *leoncillo*. Its body is elongated and its legs are very short. The ears are small and rounded while the tail is very long. Al-together these characteristics give the jaguarundi an almost weasel- or mink-like appearance. No wonder the German name is *wieselkatze*, or weasel cat.

Quite unlike the other tropical cats, the jaguarundi's fur is without pattern. It is short and smooth, not of the quality furriers seek. It occurs in several different color phases from gray or tan to black or reddish brown. Newborn kittens are spotted, as are cougar kittens, but the spots soon vanish and are replaced by the uni-colored adult coats.

Jaguarundis are neither as rare nor as secretive as margays, but nevertheless are infrequently seen. This unique species lives in dense bush, brush, and in forest edges. When seen, jaguarundi have been traveling across open or grassy areas, though never far from cover. The long, low-slung body reveals that it is mainly terrestrial, capable of movement through the thickest underbrush. Probably less nocturnal than the other wild cats sharing its Central American range, the jaguarundi is sighted most often very early and very late in the day. It is probably solitary, except for brief pairings during mating periods.

Some biologists believe that jaguarundis were fairly common a century ago along the Rio Grande River

A jaguarundi at bay could do a job on the hand that attempted to fetch it.

where it forms the boundary between Texas and Tamaulipas state, Mexico. Jaguarundis may still survive there in small numbers because sightings are regularly reported, especially in the vicinity of Santa Ana National Wildlife Refuge. But many of those sightings could be of feral domestic cats.

Early in the 20th century, naturalist F. B. Armstrong wrote the following from the Rio Grande delta regions:

"Yaguarundi cats inhabit the densest thickets where the timber (mesquite) is not very high but the underbrush (catsclaw and granjeno) is very thick and impenetrable for any large-sized animals. Their food is mice, rats, birds, and rabbits. Their slender bodies and agile movements enable them to capture their prey in the thickest of places. They climb trees, as I have shot them out of trees at night by shining their eyes while deer hunting. I capture them by burying traps at intervals along the trails that run through these thick places. I don't think they have any regular time for breeding, as I have seen young in both summer and winter, born probably in August and March. They may move around a good deal in daytime, as I have often seen them come down to a pond to drink at midday."

Farther south in Yucatan, G. F. Gaumer observed in the 1920s that the jaguarundi breeding season occurred in November or December and was marked by much noisy caterwauling and sounds of "fighting." He found litters of young, usually two, in hollow trees in January and February. Gaumer also observed that unspotted gray or reddish kittens could be born in the same litter.

Unfortunately almost all of the type of habitat F. B. Armstrong described in the Rio Grande bottomlands is now gone. The dense brush has been ruthlessly cleared for agriculture, and seasonal floods—which grow more serious each year as overgrazing occurs upstream—have wiped out the environment a jaguarundi needs to thrive. Any remaining animals are simply stragglers.

Where proper habitat does remain south of the U.S. border, jaguarundis are still skillful hunters of birds, especially of ground-nesting species. They prey often on such gallinaceous birds as quail and turkeys, chachalacas, guans, and tinamous.

During a trip the length of Central America in the 1960s, I found a jaguarundi living as a family house cat near Lake Atitlan in the Guatemalan highlands. This one, a dark brown female, seemed totally indifferent to the people around it, familiar or strange. But the cat absolutely hated dogs. Whenever a neighbor's dog came anywhere near, the cat would grimace and hiss a warning before suddenly disappearing. This semi-tame jaguarundi's life came to an abrupt end when it killed one too many of a neighbor's chickens.

During the 1960s the great game caller Murry Burnham of Marble Falls, Texas, traveled widely in Tamaulipas, Mexico. Using homemade mouth calls, he was easily able to attract ocelots and margays as well as jaguarundis into very close range.

Naturalist J. R. Rengger had ample opportunity to observe the habits of this weasel cat in Paraguay in the early 20th century, as the following account attests:

As jaguarundis only too often approach human dwellings in order to steal chickens and ducks, I often had opportunities to observe it on its forays, or to create those opportunities myself by tying a hen to a long piece of string near the hedge I knew to be inhabited by one of these predators. After a short while, the jaguarundi's head poked out from among the bromeliads, now here, now there, the animal peering around attentively in order to make sure the coast was clear. It then tried to approach my hen unseen, keeping its body close to the ground, and hardly disturbing the blades of grass in its stalk.

Having got to within six or eight feet of its victim, the jaguarundi slightly contracted its body and darted at the bird in a well-aimed spring, grabbing it by the head or neck with its teeth and immediately attempting to drag it into the cover. In stalking, it did not twitch its tail as much as other cats I have observed. It is often seen near houses, watching poultry from the cover of some shrubbery. I have never found it in trees, except when chased by dogs, in which case it passes from one tree to another with great agility. I have, however, been told that at night it takes sleeping fowl out of trees. It does not kill more than one animal at a time. Should its prey be too small to satisfy its hunger, it will immediately hunt again. On its hunting trips it does not roam very far. It swims well but does not enter the water without necessity.

# PART TWO
# THE WILD DOGS

Citizens of central Connecticut may be surprised to learn that wolves made life miserable for early residents of the region. According to yellowed old accounts, one pack led by a large old female ravaged livestock on remote farms over a wide area near Pomfret. During just one night the pack was blamed for killing 75 sheep and goats belonging to a young farmer, Israel Putnam, who would later become a daring guerrilla general in the Revolutionary War.

Putnam organized a posse of his neighbors, and they were eventually able to eliminate most of the pack. They also managed to track the female to a den from which neither dogs, gunpowder, nor fire could smoke her out. Undismayed, Putnam crawled into the hole with only a pitch-pine torch until he could see the animal's eyes and thus pinpoint its location. The farmer then retreated, loaded his shotgun, and squirmed back into the darkness again. His friends heard a muffled shot and Putnam finally emerged dragging a wolf carcass behind him. That was probably the first recorded shot fired in man's unrelenting war against wild dogs in America, a war that continues unabated to the present day.

What we have here is a paradox. On the one hand

we call the dog "man's best friend." It is our ally on hunting trips, playmate of our children, and protector of our home. A tombstone in a dog cemetery in Alabama bears this epitaph: "I loved this faithful old hound better than my wife, my children, or my country." Yet throughout history man has feared and despised his dog's wild relatives.

Though wild dogs in the New World have often feasted on farm poultry and ranch beef or mutton, there is no record that one ever attacked a man. From the werewolves of folklore to the wolf that ate Little Red Riding Hood's grandma, wild dogs have had few true friends, and that is unfortunate. For the wolves, coyotes, and foxes that make up the North American branch of the worldwide dog family, *Canidae*, are among the most interesting and admirable creatures that share our planet.

The wild dogs also share common physical and behavioral characteristics. All possess keen senses of smell, sight, and hearing. They have long, narrow muzzles, erect ears, bushy tails, and a generally slender build. In fact they may seem more fragile, less robust, than they are. All wild canines regulate body temperature by panting—cooling by the evaporation of moisture from open mouths and lolling tongues. And all evolved from common ancestors, *Miacidae*, which were tree climbing mammals that lived about 50 million years ago.

Although many kinds of wild dogs evolved, only the fittest survived into the 20th century. One North American dog now extinct is the dire wolf; all that remains is a skeleton found in the LaBrea tar pit in downtown Los Angeles. When North America was first invaded and settled by humans migrating from Asia across the then existing land bridge to Alaska, they brought "domesticated" dogs with them. These early peoples were probably also able to tame the North American wolves they found. The domesticated dogs interbred with the wolves. Later, the Spanish explorer Francisco Coronado found Indians using large mongrel dogs as beasts of burden and as sentries when, in 1541, he wandered across what is now Kansas.

All wild dogs exhibit traits similar to those that endear their domesticated relative to humans. They hold their tails erect when playing and wag them to express pleasure or to seek acceptance. They also tuck them between their hind legs when frightened or uncertain. And, like some pets, all wild dogs gorge themselves when food is available, often to stupefaction. Yet only a decade after they arrived in Massachusetts, the Pilgrims declared a bounty on wolves.

By contrast the Indians of New England had managed to get along well with wolves for centuries. Since then, millions of dollars have been spent from Florida to Alaska to eradicate wild dogs. Because coyotes and foxes are more abundant now than ever before, paying bounties is obviously no solution and is furthermore an invitation to fraud.

An interesting account of bounty fraud in 1800 comes to us from one William Cooper, a pioneer in then unsettled western New York State. One day a neighbor of Cooper's found a female wolf with six pups in a pitfall trap he had built. Leaving the family alive, he rode immediately to town and there delivered a fiery oration against wolves. He proposed doubling the bounty on the "fearsome predators" right there.

As it would be forever after, such rabble rousing was popular with local politicians and the bounty was indeed raised immediately. The man then returned to his trap, killed the wolves, and presented the skins for a higher payment. Today the same scam might be described as free enterprise.

There may be no more apt symbol of the Ameri- ▷ can wilderness than the gray or timber wolf, which once roamed widely over North America. Gray wolves were absent only from deserts, rain forests, and the highest parts of mountain ranges.

# CHAPTER SIX
# THE GRAY WOLF

I vividly recall an early morning in 1960 on the American River, a clear and slow-moving salmon stream just beyond the boundary of Katmai National Park (then Katmai Monument). I was camped with a party of Alaska Department of Game and Fish fisheries biologists near where they had built a weir across the cold, swift current. A weir is a V-shaped barrier that permits counting salmon migrating upstream, and with this information accurately managing the resource. Such weirs are often magnets to any brown bears in the area because the animals can easily catch a few of the fish that are trapped overnight.

From time to time we heard and saw the bears, which was not surprising. Their pawprints were etched in soft soil on the riverbank and all around our tent. What surprised us were the wolves. A family of five also frequented the area. Often in daylight I saw two

or three at a time splashing in the shallow water, trying to capture the fish. They were not nearly as good fishermen as the bruins, and I never actually saw one capture a salmon of its own.

Apparently the wolves had enough success to keep them in the area. Not only did they become tolerant of our presence, but two of the younger ones even seemed curious. A white wolf would sit for hours directly across from our tent, half hidden in streamside vegetation, staring at us. One morning it came to our side of the river, intently watching the tent as I unzipped the flap and stepped outside. It seemed puzzled by the aroma of coffee brewing.

I have seen and photographed wolves a number of times since that summer at Katmai, even at much closer range. But never had I been around a family for so long or so intimately. Nor had I ever heard of

◁ After having fed on the carcass of a road-killed deer, this male wolf in Montana takes his ease.

This young female wolf is playing with and gnawing on the discarded antlers of a deer.

a wolf "pack" becoming so confiding. I wrote about the experience in *Outdoor Life* magazine in January 1968. Soon after, I learned that this group of wolves—far enough from human habitation to be no threat to men or their livestock—had been shot during the winter by men illegally hunting from ski-equipped aircraft.

Exactly what kind of animal was it that would never fish for salmon again in the American River?

**RANGE AND SUBSPECIES.** Two centuries ago *Canis lupus,* the gray or timber wolf, was more widely distributed than any other mammal of historic record. Gray wolves lived across vast areas and many kinds of habitats in North America, Europe, and Asia. They were absent only from deserts, tropical rain forests, and the highest parts of lofty mountain ranges. Wolves still live in large portions of the northern hemisphere, but only in bleak and lonely northern wildernesses where humans have not yet settled in numbers. More than half their original range has been usurped by civilization and by civilized man in his efforts to exterminate them.

In North America, wolves have been completely eradicated from the entire United States except for token populations in Minnesota, Michigan's Isle Royale National Park, Alaska, and recently Montana. The wolf is also gone from the heavily populated band across southern Canada and probably from all of Mexico.

The gray wolves of this continent were divided into 23 subspecies. The most abundant of the subspecies remaining are *Canis lupus occidentalis* of Canada's Northwest Territories, *C. l. lycaon* of eastern Canada, and *C. l. hudsonicus* of the Hudson Bay region and north central Canada. Certain subspecies such as the prairie wolf, which once subsisted on the great bison herds, survives (if at all) only in captive collections. Only taxonomists are able to distinguish the various subspecies, and then only after clinical examination. Go to the wolf pens in any zoo and the comment you hear most often is that they look just like dogs. It's true of course; dogs are descendants of wolves and will interbreed with them.

Missing for many years from Montana, the howling of wolves can again be heard occasionally near Glacier National Park because a few animals have moved down from Canada.

Wolves come in many color variations, from
almost pure white to dark gray or almost black, as
shown above.

**THE PHYSICAL SPECIMEN.** It is nearly impossible to describe a typical or average wolf. Those of extreme northern Alaska, Arctic Canada, and especially the larger Arctic islands may appear snow-white at a distance. But an Arctic-white pelage close up reveals an intermingling of gray, black, or reddish-brown shades with the white. Farther south in Canada or Alaska, body colors become gradually darker. A single pack may include animals that are black, shades of brown, and nearly white. The pack I observed on the American River included all of these. Timber wolves of heavily forested parts of eastern North America tend to be more uniform in color, similar to the grizzled gray-brown found in German shepherd dogs. No matter what color it is, a wild wolf has a ghostly appearance as it drifts silently over the landscape.

The wolf's subtle color variation is a fine example of the natural selection that enables those animals best suited to their environment to survive. In the far Arctic where snow may cover the ground much of every year, a light pelage is a distinct advantage. That is why the nearly white wolf has persisted there. In the shadows and deeply mottled gray, green, and brown of eastern forests, the wolf's normal brownish

This wolf has a grizzled gray-brown coat typical of
wolves in heavily timbered eastern forests.

coat is a much better camouflage than white would be. Almost anywhere you find them, wolves blend fairly well into their backgrounds, a great benefit when they are hunting wary prey.

All wolves, and particularly those of the farthest north, have very dense underfur that insulates them against extreme winter temperatures. Still another adaptation to environment is the species' trait of hunting in packs of families or other groups. This enables wolves to kill larger animals—including any of the horned and antlered mammals that share their land—than they could alone.

**HEARING THEM HOWL.** On a cold, clear September night in 1955 I first heard the haunting, savage "symphony" that may be the most awesome sound in the outdoor world. I was camped beside a fork of the Spatzizi River in the British Columbia Cassiars with brothers Frank and Homer Sayers. We were on a Stone sheep hunt. After dinner when our campfire had burned down to dark red coals, we hobbled our horses and soon fell sound asleep in down-filled bags unrolled on the ground. I was awakened first by the brass bell on a pack horse that stomped nervously too

nearby. I sat up suddenly and heard the long, mournful howl of a wolf not far away. A cold chill actually did race down my spine.

I will never forget the chorus of chords and dischords, notes played back and forth that rolled over the alpine terrain. It seemed to be variations on a theme of some shared tragedy. The animals moaned, whined, barked, and wailed all at the same time. The alternating harmony and dissonance made me understand all the fear that has always surrounded wolves, the dread expressed by our ancestors in folklore and song. I checked to be sure my rifle was nearby.

At daybreak, before rekindling the fire for cooking breakfast, we rounded up the horses and were pleasantly surprised to find none were missing. A local Indian guide, Charlie Abou, had told me that more than once wolves had taken some of his own pack stock. During our next three weeks in the Cassiar Mountains we never again saw or heard the pack of wolves. Nor did we ever find much sign of them.

**THE PACK.** Wolves are often believed capable of selecting any prey, running it down, and killing it with impunity. This simply isn't true. The wolf's actual

Wolf society usually revolves around the pack's dominant, or alpha, male and female. They alone breed, and the "aunts and uncles" help feed and protect the dominant pair's litter.

Early spring is a busy time in a wolf pack, or family. Not yet old enough to breed, this young one keeps busy hunting jackrabbits.

size, usually from 60 to 120 pounds, is as grossly exaggerated as the size of its packs. Most scientists who have studied wolves report that packs seldom have more than twenty members and usually fewer than ten. Of course that monster image helps explain why most of the wolves of the United States were already destroyed before the 20th century.

Wolves are most gregarious. They not only hunt in packs but live their entire lives with other wolves. The trait of pack hunting has resulted in the development of complex patterns of social behavior. From studies in Canada and Alaska we know that a family—male, female, and pups—is the normal basic unit of the pack. How several other adults fit into it is still not entirely understood, however. They could be pups of previous years, or possibly second-year or adolescent animals from other disrupted packs. Or they might simply be strays that found a foster family. Adolescents are those that have been learning to hunt for a year or more and can probably pursue and help kill big-game animals with the rest of the pack.

Numerous studies of wolf packs have been made both in the wild and in captivity. Captive animals are easier to observe, of course. Both groups reveal that the highly organized social structure centers on a dominant male and dominant female. Any dominant wolf acts the part. It holds its tail high, stands stiff-legged, and bristles its mane. In a dominant presence, subservient wolves show respect by cowering with ears back, or by maintaining a slinky posture with tail between hind legs. Pack leaders achieve their rank by age, strength, experience, breeding, and determination. They are rarely challenged by others in the

group, and wolf pack members seldom fight—or *have* to fight—among themselves.

The pack bond seems strongest in winter when the wolves hunt and travel together, often for long distances. During summer the adults go on much shorter hunting forays, often for such smaller game as hares, beavers, or the young or newly born of ungulates. A few adults may hunt together, or a single adult hunts alone within range of the den or the site where pups are being raised. That site is the focus of frequent meetings, arrivals, and departures during the few months from whelping in spring until young wolves are able to follow the pack.

## MATING, BIRTH, AND DEVELOPMENT.

Wolves and all wild dogs differ from domestic dogs in their reproductive cycles. Male house pets can breed at any time; females every six months or so. Male and female wolves living wild breed only once a year. Though in captivity male wolves have bred successfully with more than one female, this probably does not happen in the wild. Whether a wild wolf mates over its entire lifetime is still being debated among biologists, but usually only the dominant pair mate. The actual breeding period varies with the latitude, but it almost always takes place during the bitter cold of March or April. Gestation is about nine weeks.

An average wolf litter contains five or six pups. But during good times—perhaps when big-game populations are high in the region, and/or when wolf numbers are low—there may be more than eight or ten pups in a den.

Across the tundra or in northern coniferous forests of Canada, dens are commonly dug in sand deposits such as eskers, which are formed by glacial melt water. In mixed forest areas the dens may be located in hollow trees, in depressions left by root masses of toppled trees, or in rock crevices. During a long-ago fishing trip to Labrador, Indians showed me a wolf den that had been vacated when those same Indians had captured the pups. The den was located beneath a great pile of crisscrossed deadfalls deposited by a river that had changed its course. A wolf pack will headquarter around a whelping den for at least a month unless it is disturbed. The disturbance is most likely to come from humans or an itinerant grizzly bear or a wandering wolverine.

A number of factors determine where wolf dens will be situated, with security and tradition among them. Some wolves den in the same place year after year. Most dens are not very far from water and many are quite close to it. But probably the most important factor is the availability of large prey species within hunting distance of the den during the five to six months when pups are confined to the den area. During that time, unless adults can bring food to the den site, the pups will starve. If pups do survive the critical first few weeks, wolves in wilderness areas will likely live to be at least five years old and may reach twice that age.

Wolf pups remain inactive inside the den for about two weeks, during which time they are fairly helpless. But then they begin to appear outside, more and more each day, playing and exploring the immediate vicinity. Social ranking of the pups may be determined by intense fighting during this early period. Often the pups will be left alone for as long as two days at a time, and then they are most vulnerable. But just as often an adult member of the pack will baby-sit while the parents are hunting. By autumn the surviving pups are traveling with the pack and beginning to participate in hunting and other pack activities.

As with most other meat-eating animals, the early rough-housing around the den helps pups to develop stamina and hunting skills. According to Canadian biologist D. H. Pimlott, mature wolves will set up ambushes or drive prey toward other wolves in the group much as human hunters ambush and drive deer.

These non-instinctive, or learned, skills are practiced in the clumsy attempts of pups to hide behind obstacles and then jump out at one another. Even in winter when they are almost fully grown, pups have been observed at play, chasing, stalking, and ambushing their siblings.

**BEHAVIORAL STUDIES.** Numerous biologists have studied wolves in North America, but none has become so totally involved with the species as L. David Mech (pronounced *Meech*). Since his first wolf investigations in 1957 on Isle Royale National Park with my friend Durward Allen of Purdue University, Mech has devoted himself to this species; and his store of knowledge is immense.

Mech figures he has flown in light aircraft for about 3,000 hours, hiked at least 2,000 miles, and covered untold thousands more miles via boat and motor vehicle in search of wolves and wolf data. He has logged enough mileage to circle the earth several times. Most of this work has been done on Isle Royale and in northern Minnesota, but Mech has also investigated wolves in Alaska and Italy, and in Manitoba and the Northwest Territories in Canada. During his first three

Sibling rivalry persists among adult wolves such as these two in Alaska and
can at times get rough.

Top photo: In the Yukon and in Alaska, Dall sheep are among the wolf's prey. But wolves and sheep have prospered on the same ranges for centuries. Caribou photo: Almost wherever herds of caribou migrate, wolves will prey upon them as they have for millennia.

The first successful hunt Mech observed began when the pack managed to separate a 300-pound, nine-month-old calf from its mother. While most of the attackers worried the cow, two chased the calf, catching it within 150 yards. One grabbed it by the rump, the other by the throat, all while running. But they couldn't bring it down until two more wolves piled on, when the calf finally fell. In minutes the moose was eaten by the entire pack. The cow moose escaped unharmed, though perhaps somewhat wiser. A study in Manitoba shows that a running moose has a poorer chance of survival than one that stands and faces the wolves that test it.

After moving his main study area to northern Minnesota, Mech found the wolves' primary prey to be whitetail deer rather than moose. Here the predators were dealing with a less powerful and pugnacious prey, yet one as alert as moose and faster afoot. Hunting success was not much better than with the larger moose, the pack usually capturing the immature or old deer, the sick or crippled, or otherwise least fit individuals.

Wherever they range, wolves primarily hunt large mammals such as elk and caribou, sheep and bison, and muskox, as well as deer and moose. They have to work extremely hard for any of this food. A summary of several wolf studies from scattered regions reveals that the animals kill only one in every nine or ten head of game they chase. In winter the victims are mostly old and weak animals, while in summer the fawns, calves, and lambs are the easiest to catch and make up most of the diet. No hungry wolf will pass up a chance to kill smaller mammals or nesting birds, however. Wolves search across waterfowl nesting grounds for ducklings and goslings. Once in Alaska I saw the carcass of a wolf whose face was a pincushion of porcupine quills.

A normally healthy wolf can survive for as long as two weeks without eating, and there is evidence they can live even longer than that. Thus they are able to hunt and perhaps chase several large mammals unsuccessfully before finally finding one they can overcome. Once wolves do make a kill, they can immediately consume great amounts of the fresh warm meat.

One morning Dave Mech and ski-plane pilot Don Murry saw 15 wolves topple a cow moose of about 600 pounds during a routine flight over the Isle Royale study area. Immediately the wolves fell upon the carcass, pulling and tearing it in all directions. Three hours later they had already eaten half of it, or about 20 pounds of meat per wolf. Although wolves lead a feast-or-famine existence in winter, Mech estimates that on the average each predator consumes five to

winters at Isle Royale, he spent about 400 hours following a pack of 15 wolves in a light ski plane. Mech devoted sixty-five of those hours to watching the pack hunt moose, their largest and most formidable prey. From those observations, he came to two conclusions.

The first was that the old Siberian proverb, "The wolf is kept fed by its feet," was true. And the second was that wolves are by no means the efficient, relentless killers they are often claimed to be. Of 131 moose that the biologist saw his wolf pack detect, the wolves took only seven. The rest escaped by their alertness, speed, endurance, and pugnacity, or some combination of these.

ten pounds of meat per day as long as hunting is good.

Biologist Harald Haugen, in a study in which he analyzed the injuries of 2,100 wolf skulls, reported definite correlation between the injuries and the size of the prey population. Where moose were scarce, wolf injuries were worse and more frequent. When moose numbers were high, injuries decreased. The difference may reflect the relative choice wolves have in their prey. With moose abundant, wolves can afford to wait for the easy victims. When moose are scarce, wolves are forced to attack a larger number of strong, healthy animals and, perhaps, thus suffer injuries. Most of the damage Haugen found was to the wolves' muzzles.

Northern wolves make plenty of footprints across the landscape, wandering far and wide to hunt. Some observers report that a pack will travel in skirmish line on a broad front. But Mech notes that, at least in winter, they normally travel tirelessly strung out in single file, at an average pace of four to five miles per hour, or twice as fast as a briskly walking human. They follow frozen waterways or windswept ridges, old game trails, and even vehicle roads. One pack Mech followed on Isle Royale covered an average of 45 miles a day between kills. Another traveled about 31 miles a day.

We find more evidence of the wolves' wandering ways in the size of an individual's or pack's range. Range is determined by following the animals in aircraft and/or by radio-collaring. Minnesota packs have been found inhabiting territories of from 50 to over 100 square miles each. Alaskan packs require larger territories.

Bob Burkholder, an Alaskan pilot-researcher, once followed a pack of ten wolves for 45 days during which the animals explored a region 100 miles long by 50 miles wide—or a territory of 5,000 square miles! The pack killed moose or caribou once each 1.7 days and averaged 24 miles between kills. Availability rather than species preference accounted for 31 total kills in the 45 days. Burkholder noted a spirit of cooperation among pack members throughout the hunt.

Dave Mech has radio-collared over 200 wolves to determine their travels. To date the record vagabond was a lone male that weighed 75 pounds when trapped in the Superior National Forest in Minnesota. Ear-tagged #1051, that wolf covered 129 miles in a straight line between the farthest points of a journey.

Mech notes that the range of loners such as #1051 may be vastly greater than of packs in the same general area. Those loners must travel carefully along the edges of established pack ranges or risk being injured or killed by their own kind. Apart from man, the animals most likely to be dangerous to wolves are wolves of other groups. In forested wilderness parts of North America where wolves are still undisturbed, all available space will be occupied by packs. Each pack has a territory of its own, overlapping neighboring territories little if at all except at times on the tundra. On those barren lands, the need to migrate to follow big-game herds seems to result in greater tolerance among the wolf groups.

Just as a family dog urinates on a fire hydrant or along the boundaries of its property in town, wolf packs mark territories and guarantee their exclusive spacing by scent marking (urinating) along the borders and by howling. Usually audible for a long distance, especially by the keen ears of other wolves, howling may be an important means to warn other packs or loners to keep away from an occupied area.

Biologists still do not fully understand all the reasons for howling. Of course it may be a means of communication among pack members as well as a warning to intruders. Some observers believe wolves howl mostly because they enjoy it, as a sort of ritual ceremony that bonds all in a pack together. Howling may also be a form of calling or coaxing to encourage stragglers, perhaps young animals, to keep up with the pack.

Still another explanation of howling is as a family roll call. During a hunt or chase individuals of a pack may become separated and distressed. Thus the howling could serve as a beacon. Just as each animal has its own odor, so each probably has a distinctive voice the others recognize. Howling may assure that all are safe and accounted for. Whatever its reason, howling is a thrilling, electrifying, even frightening sound to human ears.

Communication among family members is not limited to howling. From soon after birth until full adulthood, wolves also bark, growl, squeak, snarl, yip, and whine. One sound I have heard a number of times is the short explosive bark, which may have many uses. In Alaska's Denali National Park I heard a female announce the arrival of a grizzly bear in her vicinity by barking well before I saw the bear myself. Other observers have reported the bark as being both a warning of intruders around a den and a decoy to attract a bear away from the vicinity of a den.

Females may whine to attract cubs out of the den. Cubs whimper when they are hungry and want to nurse. Subordinate adults whimper when approaching or greeting a dominant animal. I've heard captive wolves growl to threaten one another. Adults have also been heard growling to keep pups from fighting too intensely, and females may growl to keep a certain pup from nursing. Usually a growl is a definite signal to maintain distance between individuals, whether it

A hungry gray wolf recycles a mule deer seriously weakened by winter.

be social distance within a pack hierarchy or physical distance.

**ROGUE WOLVES.** Perhaps no species of North American mammal except the grizzly bear has inspired so many tales as the gray wolf. These stories range from true or almost true to fanciful and outright fabrication. Like some of the famous old cattle- and man-killing grizzlies, rogue wolves did exist, particularly when settlers were introducing the sheep and goats that have since destroyed western range lands by overgrazing. When the natural prey animals were forced out, wolves predictably switched to mutton. Not nearly as nutritious as deer or antelope, sheep were a lot easier to catch and kill. For a brief period early in the 20th century before the sheep ranchers could get organized, wolves made livestock raising difficult in Oklahoma and Texas.

Several of the more notorious wolves of Oklahoma were made famous in the book *Animal Outlaws*, written by a state senator, Sid Graham. One lobo called Geronimo ranged over McCurtain County for years, devouring mutton wholesale before being run to exhaustion and captured alive near Broken Bow. Another gray wolf known as Lucifer was eventually brought to bay by a pack of dogs near the general store in Inola. Among other rogue wolves were the King of Cedar Canyon that roamed the Cherokee Nation, and Osage Phantom, a Tulsa County killer wolf that was never caught.

When guns and traps did not deter the livestock killers, some Texas stockmen imported packs of hounds to run the wolves as well as mountain lions and bobcats. But most ordinary hounds were no match for the faster, more durable wolves. Thus Robert Real of the Live Oak Ranch in Real County is said to have spent a fortune in buying the best bred Walker hounds in all of Kentucky. Using the hounds and widespread trapping, Real managed to kill plenty of wolves.

As late as 1912, a notorious, elusive pack of wolves still ruled the Texas Hill Country south of Kerrville, until rancher Q. T. Stevens acquired and bred his own pack of "wolf hounds" for speed and distance. He managed to wipe out the wolves, one at a time, after day-long chases as far as 25 miles on horseback. More than once he wore out his own mount and had to continue the hunt on foot. Joe T. Stevens, a Texas wildlife biologist and descendant, describes one incident:

After a 12-hour chase, Q. T. shot and skinned a large wolf. He carried it five miles back to a horse

trap where a five-year-old unbroken mare was finally corraled and saddled with a great deal of dif-
ficulty. She furiously resisted his getting into the
saddle, much less putting the wolf skin behind him,
but he managed. And that mare was about broke
by the time he rode her the remaining 20 miles
home.

The most famous wolf hunt in Texas history began
in 1905, when President Theodore Roosevelt arrived
in Dallas and from there boarded a special train to
Quanah in Hardeman County. From this base the old
Rough Rider went wolf hunting. He was accompanied
by a troop of cavalry soldiers and curious cowboys from
all over the area, as well as a party of assorted dig-
nitaries that included the Comanche chief, Quanah
Parker. Roosevelt later wrote: "We did not see any
of the so-called buffalo or timber wolves, which I
hunted in the old days on the Northern Cattle plains
[in North Dakota]. Big wolves are still found in both
Texas and Oklahoma, but they are rare now compared
to coyotes and are great wanderers."

What Roosevelt did not know, or would not admit,
was that wolves were practically gone from that entire
part of America forever.

A few wolves still clung to existence in Arizona in
the first decade of the 20th century. In fact two were
locally infamous. According to a lurid account in *Arizona Cowboy* magazine, "Old One-Toe would rip the
side out of a steer and cripple the poor beast while
he fed on it for a couple of days until it died." One-
Toe had lost the other toes of one foot in a trap,
leaving a track like "the scratch of a one-penny nail"
as it wandered and killed far and wide on both sides
of the Mexican border.

It is even harder to believe the still-repeated accounts of Old Aquila, best-known of Arizona's outlaw
wolves. For over eight years Aquila was credited with
bringing down tens of thousands of dollars' worth of
sheep and cattle, and as many as 65 sheep in just a
single night. He was said to weigh about 150 pounds
and had a price of $500 on his head, dead or alive.
But the best of hunters and trappers couldn't get close
to him. Although almost nobody ever saw the animal,
residents of the Sonoran Desert really believed that
a whole band of coyotes followed Aquila everywhere
like the tail of a comet, as reported in an old news-
paper account, "just gettin' fat on Old Aquila's leav-
ings."

Nobody knows what finally happened to Old
Aquila. Southern Arizonans simply realized one day
that the killing had stopped. By 1976, according to
the annual report of the U.S. Biological Survey for
that year, there were no more wolves left inside Ar-

izona. But just possibly, Aquila's genes survive today
among the few Mexican wolves that may still live in
the Sierra Madre Mountains of Durango and Chihuahua, northern Mexico. But most biologists familiar
with the situation do not believe any wolves still roam
there or anywhere else in the wild Southwest on either
side of the border.

**THE FUTURE.** The story of the last Mexican wolves
is interesting, though sad. As late as the 1950s a small
population of wolves, subspecies *Canis lupus baileyi*,
still skulked and howled in the highest, most remote
parts of the Sierra Madre. But as ranchers began to
graze cattle deeper and deeper into the high country,
more and more were killed by wolves. Mexican cattlemen soon began a program of extermination, assisted and advised by American hunter-trapper Roy
McBride, whom I mentioned in Chapter 1 as the cap-
turer of Florida cougars.

McBride showed the Mexicans all the lethal tricks
so successfully used against predators in the United
States, including steel traps, cyanide guns, strychnine,
and 1080 (sodium monofluoroacetate). In 1980
McBride estimated that, at most, 50 gray wolves survived in all of Mexico. All of them were isolated in
habitat rapidly being overgrazed by livestock, from
which most of their natural prey had been eliminated.

At a 1979 gathering of U.S. and Mexican conservation bureau representatives, joined by delegates from
major zoos and wildlife survival centers, it was decided
that the only way to save the Mexican wolf from extinction was to quickly trap all of the remaining wild
individuals and keep the species alive in captivity.
For the Mexican wolf it seemed to be a losing prop-
osition either way.

Curiously, Roy McBride volunteered to participate
in the unique multi-national program. No living man
was more familiar with the critically rare animal. And
McBride himself belongs to a nearly vanished breed
of tough, determined woodsmen, able to outsmart and
outlast the predator he hunts. I doubt if anyone else
would have been able to take any of the last Mexican
wolves alive and bring them to the Arizona-Sonoran
Desert Museum of Tucson.

The program has had its ups and downs, but
McBride has done his part. Up to a dozen of the animals now live in the Tucson facility and in a few
other zoos. But breeding has not been very successful,

Next page: The gray wolf subspecies *Canis lupus
baileyi* was hunted, trapped, and poisoned to low
numbers in Mexico's Sierra Madre and now exists
only in captivity.

There is no need for this Mexican gray wolf to hunt for food anymore. Food is delivered daily in this compound. Brethren of this wolf were eradicated from the wilds of northern Mexico or removed to confinement in the 1980s.

Magic Pack, moved southward from Alberta into Glacier Park where they seem to have settled permanently. They are the first group to take up residence in the western United States since the 1930s. The only other wild wolves south of the Canadian border are the 1,200 or so inhabiting northern wilderness portions of Minnesota and Michigan. The Glacier pack is making regular kills and eating well. They are even seen at frequent intervals.

The Magic Pack's arrival in Montana has raised many new questions and stimulated new as well as old, angry arguments. What if the pack divides and expands to other places in the region after a 50-year absence? What effect will it have on existing big-game populations? Do we really need another predator in an area where the grizzly bear has too often been in the news?

Yellowstone National Park, once the natural domain of gray wolves, has for several years been at the center of sentiment both for and against reestablishing wolves in the park. There are excellent reasons for bringing back the wolves. To begin, the species simply belongs there, and any visit to the park would be richer simply because these splendid natives are present. In addition, overpopulation of both elk and bison in Yellowstone today is seriously damaging much of the range. The successful introduction of wolves could only improve this situation.

Opposition to the wolves comes from three expected sources: from the cattle industry, from too many elected politicians, and from a public poorly informed by those stockmen and legislative officials.

Cattlemen claim to fear that wolves might travel beyond park boundaries (as they might indeed on occasion) and prey on livestock. But dealing with individual wolves would take care of that. Late in 1985, Senator Steve Symms of Idaho was asked about restoring wolves to Yellowstone; he commented only that wolves would "make nice carpets." Senator Alan Simpson of Wyoming stated that "as long as there are wolves in Canada and Alaska, we don't need any in Wyoming." He might just as well have said that as long as there is pure air and clean water *somewhere*, we don't need any near home.

Despite the "wisdom" of officials who should know better, wolves belong in any North American wilderness that was once their home—that is, if that wilderness still survives relatively intact. The outdoors isn't the same without them.

mostly because of the shortage of females in the captive population. Only time will tell what the outcome will be.

For decades biologists pondered why the wolves that live in southern Alberta did not cross the border into the northwestern United States, especially into the sanctuaries of Montana's Glacier National Park and adjacent Bob Marshall Wilderness. In both places the wolves would be safe from hunting. They also contained good herds of deer, elk, and moose. Wolves had lived in Glacier and the northern Rockies halfway through the 20th century; since then ecological conditions have changed very little.

Then in late 1985 a pack of 12 wolves, dubbed the

Red wolves once lived from Florida northward to ▷ the Ohio River and west to Oklahoma. Today they are one of the most endangered mammals in North America.

# CHAPTER SEVEN
# THE RED WOLF

Red wolves survived in east Texas near the Galveston Bay into the 1970s.
Efforts are underway to reintroduce them into once native regions.

As I write this chapter, four mated pairs of the red wolf, a rare wild dog, are being fitted with radio collars and introduced to a "new" home on the Alligator River National Wildlife Refuge, North Carolina. The place is best known for Kill Devil Hills on the Outer Banks where Orville and Wilbur Wright flew the first airplane.

These eight wolves, in addition to others that may be added in the next year or two, represent the reintroduction of a native species that probably vanished from the wild in the 1970s. (A few might survive in the east Texas brush country near Galveston Bay, but that is doubtful.) About 70-odd captive reds, whelped by 14 wild ones live-trapped in the coastal marshes of Louisiana and Texas, now live at the Fort Defiance Zoo in Tacoma, Washington, and in Missouri. No wonder few American outdoorsmen realize that this creature, *Canis rufus*, ever existed at all. Today it is a toss-up whether the red wolf or the black-footed ferret is the most endangered mammal on the continent.

**THE PHYSICAL SPECIMEN.** Red wolves are slender, spindly, and stiff-legged, adapted for hunting in humid southern forests and marshes. They have big ears and short coats. In appearance they resemble coyotes, though in behavior they are more like gray wolves. There may be a few people alive today who have heard the howls of both red and gray wolves, as well as coyotes. These privileged few would probably put the red wolf's voice halfway between the other two in pitch. Reds scout in pairs or small packs unlike

Weighing 50 to 60 pounds fully grown, the red wolf is lank and quick. It can live on virtually the same fare as the coyote.

coyotes, which, as I have pointed out, have lost the pack instinct in favor of individual stealth in efficiently catching small game.

An adult red wolf weighs 50 to 60 pounds, also halfway in weight between coyote and gray wolf. As lanky and quick as it is, a red wolf can live on a coyote's fare of swamp rabbits and cotton rats. Generally this highly nocturnal species needs a smaller territory to maintain itself than does the larger gray wolf, and a little more room than a coyote. The famous naturalist, hunter, and painter John James Audubon, who spent much time prowling in old red-wolf range, considered the animal to be simply a small subspecies of *Canis lupus*.

**RANGE.** Red wolves originally ranged from Florida north to the Ohio River and in a wide band westward across the Ozarks and Oklahoma, as far as central Texas. They may also have been fairly abundant in portions of the Southeast, especially in Louisiana and the Carolinas. In 1743 naturalist Mark Catesby wrote: "Wolves go in droves by night and hunt deer like hounds with dismal yelling cries." The red wolves were also convenient scapegoats for the early settlers who had to face a multitude of natural catastrophes such as floods and foul weather, mosquitos and disease, and wildlife that preyed on livestock. Though the wolves were the least of the those threats, they were singled out as the culprits. By about 1920, they had been eliminated everywhere east of the Mississippi River.

West of the Mississippi heavy timber cutting and clearing, coupled with intensive predator control, began to break up the centuries-old red-wolf society. Unlike coyotes, the reds proved extremely easy to trap. As soon as shattered red-wolf packs stopped defending their territories, coyotes moved in. Occasionally they interbred with surviving wolves, but mostly the coyotes just replaced these original residents.

In areas of southeastern Texas and southwestern Louisiana—the last stronghold where a few red wolves held on longest—the wolves and coyotes mated indiscriminately, creating what biologists referred to as a *hybrid swarm*. The genetic identity of the red wolf was gradually lost as the wild dogs more and more resembled coyotes. Only in the early 1970s was a belated effort to save the last of the red wolves begun.

**TRANSLOCATION.** The project, based in Beaumont, Texas, was headed by federal biologist Curtis Carley. It consisted mostly of trapping the wild dogs in modified steel traps to which tranquilizer tabs had been attached. Most captured animals tended to gnaw at the traps and thereby became sedated before they could chew off a paw or otherwise injure themselves.

Most of the animals caught were coyotes or hybrids, but a few were pure red wolves. The main problem was telling the wolflike hybrids from the true wolves. Final determination was made by measuring size and weight, and mostly by X-raying the skulls to determine the ratio between brain volume and skull.

After careful examination for disease and an innoculation, some radio-collared purebreds were released in the same area from which they had come in an effort to study their movements and habits. Most of the others were sent to the official red-wolf captive breeding center at the Fort Defiance Zoo. In 1978 Carley estimated that only about 50 red wolves remained in Texas and Louisiana, and predicted that all would be hybridized within the following year or two. That has come to pass. Although the captive breeding has maintained the wolf's gene pool, Carley concluded that the only hope for the species to remain pure in the wild was to reestablish them into their former range at a place where coyotes did not already exist.

Bull Island on the Cape Romain National Wildlife Refuge was chosen as the site for the first attempt at translocation. The 7,500-acre barrier island lies just

Beach areas of the Gulf of Mexico coast were also inhabited by red wolves, as they were here at Horn Island, Mississippi.

Red wolves of southern forests originally shared much of their range with wild turkeys and, no doubt, preyed especially on the poults.

offshore between Charleston and Georgetown, South Carolina, and public access is limited. I hunted whitetail deer there during an archery season long ago and found it to be an extremely beautiful strip of coastal real estate. Potential red-wolf prey species such as raccoons and deer, fox and squirrels, swamp rabbits and quail seemed to be numerous. Overbrowsing of vegetation indicated a deer herd too large for its range, making the introduction of large predators a probable blessing.

The first pair of wolves that had been captured near Beaumont in January 1976 were flown to Bull Island eleven months later. They were confined to a pen and fed a diet of native prey for a month to help them adjust to their new environment. Radio-collared, the pair were released and tracked by biologists in a jeep carrying telemetry equipment.

For a week or so all seemed to go well as the wolves explored their new home. But things began to go wrong when the female unaccountably crossed a three-mile tidal marsh to the South Carolina mainland in broad daylight. That was especially strange because red wolves travel mostly at night. The female was trapped and returned to the pen on Bull Island, but in April 1977 the wolf died of a uterine infection. Another pair of released wolves remained on Bull Island for a year, then vanished, and the site was abandoned. But the effort did give valuable information about handling and transporting the rare wild dogs.

The next site selected for a new permanent home was the 170,000-acre recreation area in western Kentucky between Kentucky and Barkley lakes, called Land Between the Lakes. A century ago red wolves lived here, but in 1983 fierce opposition from hunters and farmers eliminated that possibility in our time. Another proposal to reintroduce wolves on South Carolina's 300-square-mile Savannah River Plant (a

federal complex that produces plutonium and tritium for nuclear weapons) was also discouraged. Thus, wolf biologists settled for South Carolina's Alligator River National Wildlife Refuge.

Although heavily timbered around the turn of the century, this peninsula—surrounded by the Alligator River, Croatan, Albermarle, and Pamlico sounds—has remained relatively untouched since then. The Refuge is mostly wild swamp and marshland mixed with dank woodland. It is home to much assorted wildlife, including bobcats and rabbits, warblers and wading birds, and even a few black bears. The peninsula is also the northernmost habitat of the alligator and any other crocodilian, as it remains protected from development by its inaccessibility and, according to one local politician, its "uselessness."

During the early 1970s First Colony Farms, an agribusiness corporation, had acquired most of the land in a joint venture with the Prudential Insurance Company. Their aim was to drain and convert it to corn and soybean production. But faced with falling land and farm commodity prices, in addition to the great cost of clearing the land, Prudential was convinced by the Nature Conservancy to spare the peninsula as a natural area. Thanks to the initial efforts of the Conservancy—an Arlington, Virginia-based national organization dedicated to preserving wild species and habitat—Prudential eventually donated 120,000 acres to the U.S. Fish & Wildlife Service as a federal refuge. The land may be the red wolf's final chance to survive in the wild, because money for such valuable conservation work is becoming almost as rare as the wolves themselves.

Alligator River is prime red-wolf habitat. To date

A last stronghold of red wolves was Anahuac National Wildlife Refuge in southeast Texas, where we found this large alligator but no wolves.

93

It is uncertain whether any of the last-remaining red wolves will thrive in areas into which they are being reintroduced.

it is free of coyotes and the possibility of interbreeding. It adjoins a 46,000-acre Air Force bombing range where most of the land is wooded and also good wolf habitat. Furthermore, the introduction of wolves has been meticulously planned.

**THE FUTURE.** In too many instances where even the possibility of introducing wild predators is mentioned, local residents oppose it with a predictable reaction. They just don't want any new "killers" running around. But this time officials are going more slowly. They seem to have won local North Carolina residents to their side. Only time will tell.

During the 1970s Peggy and I made two trips to Chambers and Jefferson counties and the Anahuac National Wildlife Refuge in extreme southeastern Texas. We hoped to see or at least hear one of the red wolves howl in their last stronghold. What we found in that part of coastal Texas was a humid forest of elm and sweet gum trees, hackberries and loblolly pines, and beech and oak, just inland from the Gulf of Mexico. Between the Gulf water's edge and the woods was a strip of grassy plain of bluestem grasses and switchgrass, gamagrass and Indian grass, punctuated by patches of milkweed and bluebells. Hiking through the knee-deep vegetation, we came upon sandy knolls overgrown with myrtle and yaupon; in times past, red wolves might have denned there in seclusion. We found the scattered tracks of wild canids, but had no way of knowing whether they were made by coyotes, red wolves, or (most likely) hybrids.

We walked along a low bayshore ridge parallel to the Gulf where both Indians and red wolves had searched for crabs and for fish washed ashore after a storm. In the spunkweed and cattail marshes of Anahuac we saw many large alligators, and waterfowl flushed whenever we drove too close along the narrow refuge roads. I'm sure the red wolves here may often have feasted on geese or ducks, especially any they could catch on their nests. The wolves probably also fed on the bullheads and gars that were stranded in shallow pools and bayous during extended dry periods. Later when settlers brought cattle into the region, the wolves no doubt fed on the bloated carcasses of animals that could not cope with the heat and humidity. Red wolves, according to tales of east-Texas old-timers, developed a fondness for the hogs that settlers introduced. Today, with almost nothing to prey on them, the escaped hogs have proliferated to such an extent that they are a serious menace to other wildlife and to the native vegetation.

On our final morning there one April, while driving slowly along a trail in the Anahuac Refuge and watching the land all around, I rounded a bend and jammed on the brakes. Trotting down the trail a hundred yards ahead was what appeared to be a wolf. It paused a split-second after I hit the brake to turn and look back at us. "Wolf!" I said to Peggy and focused binoculars on the animal.

The dull reddish-brown color seemed right for a wolf. The animal even seemed to have the lighter cinnamon color around the muzzle and eyes. But as I studied the animal, I realized it had the foxlike head of a coyote rather than the heavier head and broader muzzle of a red wolf. And when it ran, the case was closed. My wolf was a coyote. Only a coyote runs exactly like a coyote.

So I suppose I will never see a red wolf in the wild. We can only hope that those captive animals being released along the Alligator River in coastal North Carolina somehow survive and live for future generations to discover.

The coyote may still be the song dog of the North ▷
and the American prairies, but this clever wild dog
has now also become well established in America
from coast to coast.

# THE COYOTE

ne evening many years ago during our first winter in Jackson Hole, Wyoming, I brewed a huge batch of stout and fragrant venison chili in a blackened old Dutch oven. After dinner I placed leftovers, still the weighty vessel, on the back porch to cool and probably freeze overnight. Hours later I heard the heavy lid being pried from the container. I got out of bed as quietly as possible and went to the window. Illuminated by the moon was a light-colored coyote gulping down our next day's dinner.

I've often wondered how that freeloader fared. I like chili hotly spiced, and this concoction was not of the mild variety. But in the morning I found the pot licked clean.

Come to think of it, that coyote may have been surviving for a long long time on victuals even stranger than that hot homemade chili. There is a strong sen-

timent in the western United States that after all human and animal life has vanished from the face of the earth, coyotes alone will somehow survive. They seem to be indestructible.

Other native carnivores such as grizzly bears, wolves, and black-footed ferrets have almost disappeared in the wake of bulldozers and plows, fences and firearms. Predator controllers have used electronics, the deadliest poisons, and even the synthetic female hormone stilbestrol in an expensive coyote birth-control campaign. Coyotes are considered vermin everywhere, shot on sight or trapped wherever a rancher finds fresh footprints. Unprotected by law, *Canis latrans*, unlike the grizzly and gray wolf, not only survives but prospers. There may be as many in America today as ever before.

As one wildlife biologist noted, the species should

◁ "The little song dog of the prairies has gone tiptoeing over the landscape . . . sampling new foods, testing new hazards, to become the central figure in the most compelling wildlife success story in the history of our continent."—*George Laycock*

Every winter, northern coyotes grow the splendid, warm, long-haired pelage needed to survive brutal weather. In such cold climates, coyotes survive largely on the carcasses of big game unable to cope with the cold and deep snow.

have been named *Canis controversia* because the war between the so-called "brush wolf" and the sheep ranchers in particular rages as loudly as ever. It seems that each week we read about some new scheme, device, or appropriation to wipe out coyotes. So how does this furtive animal cope with civilization?

According to nature writer George Laycock, the coyote possesses an uncanny sensitivity to threats to its welfare. With an inherent intelligence and broad tastes in food " . . . the little song dog of the prairies has gone tiptoeing over the landscape, exploring new places, sampling new foods, testing new hazards, to become the central figure in the most compelling wildlife success story in the history of our continent."

## RANGE AND PHYSICAL CHARACTERISTICS.

What is most amazing, probably, is that during the period of heaviest persecution the coyote not only maintained its numbers in western North America, but has greatly expanded its territory in all directions. Today an estimated 2,000 to 3,000 live permanently within Los Angeles city limits from Laguna Beach to the Hollywood Hills. They drink the chlorinated water in expensive swimming pools, prey on house cats, and eat bowls of vitamin-fortified dog food left at back doors for pampered family pets. Coyotes are now marching southward through Central America, and to the north and east they have reached the coasts of Arctic Alaska and the Atlantic seaboard. No doubt this expansion is one result of undue disturbance elsewhere. Although the species can now be seen almost anywhere, in my mind it really belongs in open country—in sagebrush and buffalo grass, along the edges of evergreen woods and "where the deer and the antelope play."

For all of its formidable abilities, the coyote is not a large or majestic mammal to match its image. During an 1861 trip overland to Nevada, Mark Twain described it as "scraggly, scruffy, and despicable in appearance, always running away in a deceitful little trot." Especially in spring and summer a coyote's coat may be on the shaggy side, but otherwise the average 25-pound adult males are as sleek and attractive as any groomed domestic dogs. Males are less than two feet high at the shoulder and measure about four feet from nose to tip of a bushy, usually drooping tail. Females are a little smaller and weigh a little less, while coyotes of the northern portions of their range average slightly larger than those in the Southwest.

Coyotes also can differ greatly in color. Like the chili-eating coyote, most of its kin that we meet in northwestern Wyoming are light-colored, pale gray to creamy, sometimes in midwinter appearing white. But the coyotes I have watched for many hours from blinds in south Texas brush country are brownish and much darker. They blend extremely well into their sere surroundings. A fur buyer who travels through Texas and points west told me that he has seen a few pelts that were nearly black.

## SONG OF THE COYOTE.

Despite their abundance and current wide range, coyotes are more often heard than seen by all but the keenest outdoor observers. They may bark or howl at any time of day, especially throughout winter, but late evenings are the best times to hear their haunting, unforgettable song. A typical rendition may begin during an autumn dusk when one animal opens with a series of sharp yips that stretch out into a lonely, mournful howl. Quickly another animal answers from a short distance and finally these are joined by coyote songs coming in from other directions. Although it may sound at first as if all are commiserating with one another, almost certainly it is instead a family communicating among themselves, with perhaps an extra coyote or two chiming in with its special messages.

Coyotes howl for a number of reasons, such as to warn of danger, to assemble the family, or to show their exuberance. These photos show coyotes howling in answer to calls of kin not far away.

This coyote has captured a squirrel.

A mule deer weakened or injured by the rutting season just past, as well as by the intense cold, has become this coyote's prime means of survival.

Coyote howling is not well understood, even by the scientists who have studied the animals the most intensively. Biologist Franz Camenzind concentrated his coyote research around the National Elk Refuge at Jackson, Wyoming, where there is a good and largely unmolested population of the species. He has identified several kinds of specific vocalizations. A single animal howling may be a summons for the group to assemble. One animal barking may be a warning of danger, as for example of a human approaching a den. And a single coyote "bark-howling" could be a warning or challenge to unassociated coyotes to keep their proper distance.

Most biologists believe that the yipping and howling of a group we most frequently hear is a reinforcement of family ties and a declaration of territorial ownership. It is not simply howling at the moon, as some old-timers and Indians believed, nor is it the celebration of a successful chase. When howling, coyotes may very well be enjoying themselves.

Though group howling isn't as readily induced as individual howling, my friend Brent Allen and his wife and son, of Columbia Falls, Montana, can together imitate a group howl so effectively that any wild coyotes within hearing will answer or join in. Tales are frequent about coyotes responding to the doleful whistle of steam engines crossing the American plains in years past. I knew of one captive coyote that would respond to every fire engine or ambulance siren passing its yard. Southwestern folklore author J. Frank Dobie once wrote about the Scottish immigrant in Texas who entertained himself on cool evenings by playing his bagpipe outdoors. Almost always the local coyotes would harmonize with the strange Highland music. Peggy and I also heard of a Texas cowpoke who could coax coyotes to sing by playing range tunes on his harmonica.

**SKILLED HUNTERS.** Many people assume the coyote is a loner because most of the time it is seen by itself. The theory is that the coyote has not evolved a society because one animal alone can catch all the rabbits and other rodents it needs. Wolves, by contrast, hunt in packs because their prey is larger and stronger. A Minnesota coyote study tends to support that theory. Sixty percent of the animals in the study group hunted and traveled alone all winter, while the other 40 percent associated with only one other coyote.

But then how can we explain the group harmonizing I described earlier? Although most of the coyotes I've seen in Wyoming were indeed traveling alone, I have also noticed groups of as many as five together, exploring the National Elk Refuge in winter. Indeed, coyotes do at times form group bonds. Friendly ones, perhaps of the same family, not only recognize but greet one another and probably also cooperate in other ways. There are many witnesses to such cooperation in hunting, as when one animal digs into the entrance of a ground squirrel burrow while another waits for the rodent to attempt an escape from the other end of the passageway.

From a vantage point on a ridge, a Montana game warden watched a pair of coyotes methodically crisscross a sagebrush flat, never overlapping paths, and finally locate a crouched antelope fawn which they killed and ate together on the spot. Another observer reported one coyote jumping up and down as if to mesmerize or hold the attention of a crouched jackrabbit while a partner coyote crept up behind the hare unseen.

Alone or in partnership, coyotes are successful hunters because they have specialized sensory equipment. Their way of life demands it. A coyote's eyes are much more sensitive than man's to details of the landscape. The coyote can see a mouse moving through grass at a range beyond the point where, for humans, the mouse merges into the background. Sounds a human could never detect betray a prey animal's position to a coyote because of the predator's large outer ears that gather the softest of sounds. Some other carnivores depend on teeth and claws together in actual killing, but coyotes use only sharp, enamel-capped canine teeth combined with modest speed afoot.

An adult coyote immobilizes its prey by grasping it by the throat. A crushed larynx results in death by suffocation. In fact, puncture wounds and contusions in the throat area are the telltale signs of a coyote kill. Mangled flanks or hindquarters, on the other hand, are usually the work of free-running domestic dogs. Such damage is routinely blamed on coyotes.

Coyotes have been compared to wise supermarket shoppers who always seek out the special of the week, be it rib steak or hamburger. Quickly and instinctively, the animal measures a potential kill in terms of energy required, season, and odds of success. Since dead animals neither flee nor fight back, carrion is always the coyote's best bargain—unless there is some suspicious smell or device about the carcass such as poison or a trap. Though a few foolish, usually younger coyotes become victims around carcasses deliberately placed as bait, wiser animals walk away and look for the next-best bargain.

Most of the time that will be a hare, ground squirrel, or rabbit, a ground-nesting or game bird, or a fawn. I have seen coyotes chasing healthy deer, and vice versa. Once in Yellowstone Park, four coyotes managed to encircle and isolate a mule deer on a steep slope just southeast of Mammoth. But night fell and I never found out how the life-and-death drama ended.

Among other observers, I have seen coyotes use the irrigation and drainage ditches that crisscross western grainfields as a trench in which to stalk geese and ducks feeding on the grains left by artificial pickers. By crouching low, a patient coyote can occasionally reach within a few feet of an unsuspecting,

Hunting with keen ears as well as excellent eyesight, a Wyoming coyote zeroes in on its prey and lunges for it.

An antelope ground squirrel, such as this one, lost in the chase shown at left.

feeding goose. Coyotes also learn to loiter around marshes popular with duck hunters every fall. There the wild dogs are able to live on crippled, unretrieved waterfowl until the hunting season ends. Coyotes have been shot and then found to have lead pellets in their stomachs from eating downed birds.

Like many other creatures in this book, coyotes are really omnivores rather than pure carnivores. I've seen them eat mesquite beans, rose hips, and chokecherries. A substantial part of a fall diet might be fruit and berries, wild or domestic, or even sweet corn; in summer, garden vegetables are added to the list. Though the animals are especially hated by sheep growers, they are probably more despised in some regions by farmers who raise cantaloupes and watermelons, for which coyotes can develop a strong appetite. During summers of grasshopper or locust infestations, they gorge on these pests.

To attract coyotes close enough for some of the photographs in this book, I have used various kinds of baits. Among those that work best are, understandably, ground squirrels that are picked up as road kills, and—not so understandably—sugary fruit preserves and peanut butter. They also like all brands of commercial dog food, the more of it the better.

**MATING AND BIRTHS.** The coyote mating season is in midwinter, which means February in Wyoming when females come into estrus. They are not strictly monogamous, but most biologists agree that adults will maintain family ties if not harassed or unduly disturbed within their territory. A paired male and female will probably remain paired until one of them dies. Breeding pairs hunt together; at least midwinter is when I have most often seen two at a time.

This pair of California coyotes were chasing an antelope ground squirrel. The lead coyote caught it, and the slower coyote glowered when its hunting partner did not share the prize.

Later on, the male will also help provide food for the young.

Gestation is about two months. As winter is slowly blending into early springtime, the female begins to prepare several dens, one of which will be the nursery. Denning habits seem to vary in different parts of America. In some regions old underground dens are often reused; elsewhere females select new sites every year, although these may be in the same general area. Most are well hidden, in places where people are not likely to intrude. But I have found at least two dens that were clearly visible. One was only 200 yards above, and within sight of, a highway near Daniel, Wyoming.

In the northern Rockies most pups are whelped during mid-April. Earlier and later whelping has been recorded in Texas and New Mexico, however. The average litter ranges from four to seven, though a litter of nine such as the one discovered on the National Elk Refuge is not uncommon. Examination of uterine scars in trapped females revealed that one had given birth to a dozen pups. Professional predator trappers report digging out 19 pups from a single den, but that was probably the result of two females sharing the site—a rare occurrence.

Among the most interesting facts concerning this interesting mammal is that females tend to have larger litters in areas where they are under greatest stress. In the Jackson Hole area where predator control pressure is minimal, litter size averages 4.5 pups. Elsewhere in the American West the average is six. This seems to suggest a natural compensation for losses from trapping, shooting and other predator control. It also helps to explain how coyotes cope so much better than other persecuted creatures.

The coyote den offers another subtle clue to the animal's ability to survive in a hostile world. Almost without exception, a coyote raised in the wild by other coyotes will remain forever wild and fearful of humans. One mistake, one human suddenly too nearby on the landscape, and its life could be over. On the other hand, a coyote pup raised from birth by humans may become moderately tame and trusting.

Coyote pups are born with eyes sealed. Their wooly dark-gray fur is soft. The mother probably does not leave them for the first few days. But soon she begins to spend more and more time outdoors, resting some distance away and usually downwind from the den, alert for danger. At about three weeks of age the pups totter and stumble upward toward the mouth of the

Only two pups comprise this coyote litter. They kept close to the home den for more than a month while the parents hunted far away.

den to emerge for the first time into daylight. They blink uncertainly in the strange light and warmth of the May sunshine. Pups gain strength quickly in their unsteady legs and soon are play-fighting, wrestling and stalking grasshoppers around the entrance to the den.

**DEN LIFE.** I once located an active den in Grand Teton National Park and spent several afternoons watching the activity through a spotting scope from a very long range. Toward dusk I saw an adult—the female, it turned out—cautiously approach the den from the opposite direction. She finally paused by the entrance and four pups scrambled out, climbing over one another to grope for the teats on her belly. Like all wild canids, the female nursed them standing up for several minutes. Then suddenly she departed again. The pups fell back inside the hole, and the den site seemed deserted.

The next day, I wanted to get closer to the den to shoot photos. I moved nearer and, when none of the occupants were in sight, set up a small portable cloth blind. But that was my mistake. During the next night the female moved all the pups to one of the alternate den sites she had selected before whelping. It took me three days to find the new den, and this time I kept my distance. More than once I've watched adults return and regurgitate food for the pups. One morning a parent carried a ground squirrel, which may have signaled the beginning of the end of nursing activity.

How might an adult coyote carry four pups for almost a mile to a new home? By the loose skin on the shoulder or middle of the back, or even by a hind leg. Apparently they are always carried one at a time.

A pup may know more than one den as home before it is weaned because a mother coyote will carry her pups to an alternate den if she feels the first has been noticed by predators.

Hope Ryden, a naturalist-writer who kept vigil on a den in the National Elk Refuge, saw a coyote awkwardly transport a pup so that its head bumped the ground throughout the trip. In his fine book *The World of the Coyote*, Joe Van Wormer writes of one adult that carried four pups a distance of five miles from one den to another, overnight. Simple math shows that the distance covered was 35 miles over uneven terrain.

Although the playful life around a den may seem idyllic to a human observer sitting far away, it is a hazardous time for the occupants. If half the litter survive to be a year old, the coyotes are doing well. Men who smoke, burn, fumigate, or dig out the pups are a threat outside of designated sanctuaries everywhere. Wildlife predators that might capture and kill pups include bears, eagles, wolverines, wolves, and the wild cats. Females may or may not try to defend their pups, depending on the situation.

At least one coyote mother we saw was extraordinarily brave—or foolish. Peggy and I were driving to Yellowstone Park one afternoon along the Snake River near the south entrance. It was May and two sibling grizzly bears, probably 2½ years old and weighing 175 to 200 pounds apiece, were grazing side by side on patches of grass where the snow had melted. Suddenly as we watched the bruins, a coyote sprang from a ground hole directly in front of them and attacked head on, biting and snarling. The attack was so sharp and so unexpected that the young grizzlies turned and ran away up a steep bank. The coyote continued the pursuit, biting both on the rumps until all were out of sight. Thus the mother saved her family in a den in the grizzlies' path, at least for the time being. She probably moved all to another, safer site during the cover of darkness that night.

Pups that survive until fall or early winter may linger as part of a family group. Or, depending on food supplies, available territory, and social pressures we do not understand, they may disperse to seek a domain of their own. Any first winter alone is bound to be hazardous. Only the strongest survive until spring. The weak are most likely victims of the same predator control programs that capture very few of the older, wiser coyotes.

Recently travelers along the Trans Canada Highway reported seeing four wolves run down a coyote on the frozen surface of a lake near Jasper, Alberta. And a Minnesota game warden saw wolves catch another coyote on the ice near Ely and quickly tear it apart. Wherever the two wild dogs inhabit the same range, the coyote is kept under pressure. In fact coyote abundance is one important consequence of eliminating wolves from their natural range.

**BADGERS.** One puzzling aspect of coyote behavior concerns the coyote's unique association with badgers. Different observers have reported that the two play, roam, and hunt together, even live together. Some early Plains Indians believed them to be of the same family. But much of the so-called badger-coyote relationship is pure myth.

According to biologist Dr. Steve Minta, who long studied badgers on the National Elk Refuge, coyotes do at times follow badgers around and seem to be cooperating with them. But they have simply learned that a badger, which is an extraordinarily powerful digger, will often flush ground squirrels from their dens. That makes the squirrels easy prey for the coyote, which is much quicker than a badger. But coyotes do not share their catches with their "partners" to maintain the partnership as some writers have claimed. It is just another case of the wild dog using all the odds in its own favor.

"They live off the fat," one west Texas rancher told me, "or actually the fat*heads* of the land."

**SUBURBAN ADAPTATION.** But while individual coyotes may be weak and soon weeded out, the species is strong and very adaptable. It is also flexible in its needs and habits. Seemingly weatherproof inside its thick fur coat, and tireless, a coyote can trot for hours on end, looking for a better deal somewhere else. With agile, supple bodies, they sprint, leap, burrow, and squeeze easily through fences. Waterways are never serious barriers to reaching a destination. One coyote was seen swimming the Potomac River near Mt. Vernon, Virginia, almost within sight of the high-rise apartments of Washington, D.C. It isn't any wonder the coyote is now being called the Song Dog of the Suburbs, though Wonder Dog might be a better description.

Coyotes are no longer an exotic species in Westchester County, a northern suburb of New York City. It isn't that unusual to see one, especially when driving the roads and parkways late at night. There coyotes are now established, probably permanently, among the corporate parks and genteel homes of Armonk, as well as on the small farms surrounding White Plains and even southeastward along the shores of Long Island Sound.

Although the arrival of coyotes in the East isn't making many easterners happy, the critters may be

When food is plentiful and the living is easier, coyotes have time for play—
probably a holdover from the sibling rivalry pups feel around the dens.

gaining a few new friends back in the West. Some of the same cattle ranchers whose knee-jerk reaction was to shoot first and justify later today have taken a fresh look at an old nemesis. They see the coyote more as something that recycles the jackrabbits and mice that compete with cows for available forage. "The only trouble with coyotes," one central Wyoming cattle rancher confided to me, "is that they don't eat enough. Of sheep, that is."

Ironic as it may seem, some of the most extravagant admirers of *Canis latrans* today are among the professionals who spent so many years in government predator-control work. Many see their past work as a job that then needed attention. But they weren't unhappy when the use of poisons such as Compound 1080 was finally outlawed from their arsenal. "It always seemed unfair to use poison on the smartest animals in the wilds," one retired trapper told me ruefully. At the time of this writing, 1080 has been re-approved in Washington.

According to another retired government man in south-central Wyoming: "We've just created a race of super-coyotes. For example, the coyotes around here won't look at any kind of carcasses any more. In some places I know they don't move until after dark, either."

**A MATCH FOR MAN?** Every veteran trapper has his own stories to tell about super-smart coyotes that outwitted him. One government agent who admitted to me that he had trapped over 2,500 coyotes (including 220 in one eight-month period) before he stopped counting told me about one nemesis. He said, "It knew more about traps than the people who manufacture them."

Time and again this trapper would place a trap along the trail used by this old coyote, bury it carefully, and sprinkle coyote urine around it to indicate that a strange animal was invading the territory. But just as often he returned only to find the trap empty, but sprung and laying exposed above ground. The coyote

had dug beneath it until the trap jaws snapped shut and the contraption was rendered harmless. Exasperated, the trapper then used a trick that never had failed before. He shot a prairie dog and stuffed it far back beneath a flat rock beside the coyote's regular trail. Then he carefully buried four traps all around the baited rock. Two days later all of the traps had been dug up and the prairie dog carcass was gone. You have to admire an adversary like that.

For many trappers, getting a certain coyote became a fascinating game, and some curious tricks and tactics were used. One man in Nebraska would stake out live house kittens among his hidden traps. Another would bury an old-style wind-up alarm clock with a trap right on top of it. Curious about the ticking they heard underground, younger or less cautious coyotes were trapped with this ruse.

Some trapped coyotes still managed to escape from leghold traps by twisting or chewing off a foot, and probably most of those survived to hunt again on three legs. In Oklahoma a female left both forefeet in a trap, but still managed to survive on hind legs alone. More remarkable still is the female in California that was healthy and the mother of a litter of pups, even though completely blind.

With rare exceptions, the coyote is currently taking care of itself very well. People are even becoming accustomed to its presence from Georgia to Maine where it is a new arrival. In some areas the coyote is regarded not as the wild dog or the song dog, but as the public dog. Recently there was a proposal in California's Los Angeles County, to start thinning out the large population of coyotes there.

"Ridiculous," responded Dennis Kroeplin, the San Fernando Valley wildlife control officer. "The coyotes are beneficial; they keep rodent populations under control, and no Los Angeles coyote has been known to attack a human. At the same time, 52,000 dog-bite cases are reported here every single year."

He had a very good point there.

The red fox is by far the most abundant and prob- ▷
ably the most resourceful of all native foxes. It
thrives both in wilderness and near civilization.

# CHAPTER NINE
# THE FOXES

At least one of the five native foxes inhabits every region of North America, including many metropolitan areas. These are the red fox, gray fox, Arctic fox, swift fox, and kit fox. The red ranges everywhere from the Mexican border northward and throughout Canada, except for a wide band in the Great Plains, in Florida, and in the Southwest. The gray fox thrives over most of the United States except for the northwestern quarter. It also roams southward throughout Central America. The range of the Arctic fox covers northernmost Canada and Alaska. Kit foxes are natives of the American Southwest and Baja California, while the swift fox is a dweller of the Great Plains of the central United States.

**THE RED FOX.** Not only is its range the most extensive, but the red fox, *Vulpes fulva,* is also by far the most abundant and certainly the most resourceful of all the foxes. In human terms it might be regarded as the ultimate materialist. Every red fox eats very well, wears a beautiful coat, and is fairly good to its family. It has a wry sense of humor and doesn't seem to care what anybody thinks. Scientists have long realized that some wild species simply collapse at the approach of civilization, while others seem to laugh at man. Rowdy Red is among the latter.

If a man is imprudent enough to grow grapes or raise poultry, to plant melons or berries, well, how can a self-respecting red fox ignore the invitation to share the bounty? Most of our cultural developments have been to the species' liking. In the 16th century Prince Nicolo Machiavelli said that "to survive, a man must know how to play the fox," hence the word "Machiavellian" for foxy behavior. Older still is the biblical Song of Solomon that mentions "the little foxes that spoil the vines."

◁ A fox's eye pupils are oval, like that of cats, closing to slits in bright sunlight and opening large at night, allowing superior night vision.

When a red fox rests during midday, it usually does so in an open place with good visibility all around as it lies curled up snugly in its own tail.

Relying on its keen hearing, a fox often hunts mice, moles, and
shrews by zeroing-in on sounds alone and then pouncing feet-first
upon its unseen prey, pinning it before digging it out.

Not all researchers agree that the red fox is really a native of North America, although it probably lived in the open country of Canada and extreme northern United States when Europeans began to colonize the vast forested areas south of the Great Lakes. Until settlement, these forests were the natural habitat of gray foxes. But during the 1700s, wealthy tobacco planters and plantation owners, anxious to carry on the fox-hunting traditions of England, found that gray foxes would not "run well" before the hounds. So they imported European red foxes from the British Isles and released them all along the Eastern Seaboard. What followed was a population explosion of the red imports wherever the land was cleared for agriculture. Eventually the Old World red foxes met the New World red foxes (which were practically the same) and interbred. What we have now is a very clever as well as cosmopolitan creature.

What exactly is this wily culprit that seems to sneer at man? It is a small canine roughly a foot high at the shoulder, weighing 8 to 10 pounds when fully grown, with a bushy tail longer than the animal is tall. This tail streams out behind when the fox runs. Altogether a very handsome critter, the fox's thick, usually red, silky coat makes the animal seem much larger than it is. Especially in northern parts of the range, red fox pelts may vary from yellowish or brassy to black—or blue or silver, depending on the angle of the sunlight. Animals of the darker color phases are called *cross* foxes. All are lighter colored on the bellies, with dark feet and white-tipped tails.

The red fox may actually furnish evidence that some wild animals have the ability to reason, at least if you believe a story that is often repeated in rural mid-America. Foxes often have fleas that in midsummer make their life very unpleasant. To get rid of them, according to accounts, a fox will carry a twig or piece of bark in its mouth and walk out into a river or pond until only its nose sticks out. To avoid drowning, the fleas find a last refuge on the twig, which the cooled and greatly relieved fox leaves afloat in the water when it returns to dry land. Even if that never really happens, as some naturalists insist, it does sound like a ploy a red fox would use.

I suspect that northern red foxes are more active during daylight, although they are well equipped for night hunting. The pupils in a fox's eyes are not rounded as in other canids, but are oval like a cat's. This larger opening allows more light to enter during darkness.

To compensate for the wear caused by gnawing and chewing, new layers are constantly added to the teeth of a red fox. Therefore a fox's age can be determined by counting tooth cross-sectional growth rings, just as a forester calculates a tree's age by counting annual rings in the trunk.

The nose and especially the ears of every red fox are exceedingly sensitive. They can hear a mouse, mole, or shrew moving under a distant mat of vegetation. Its pointed, furry ears also can detect the squeak of a rodent 50 feet or more away. Typical hunting strategy is to pause motionless in a fallow cropfield with one paw raised, listening intently. When its senses zero in on a mouse, the red fox rears up on hind legs and pounces with back arched, pinning the prey, which it has not yet seen, to the ground. If the fox is in a hurry, the mouse is gulped down whole. But once from a goose-hunting blind in Maryland, I watched an adult fox play for several minutes with a mouse before finally eating it.

Early in my life I spent too many days and nights hunting foxes with hound dogs in the hills of Lawrence County, southern Ohio. It was mostly wasted effort because during that time I can recall capturing only two foxes. Mostly, even with the finest dogs, it was like tracking a ghost over ridges of blackberry briers and laurel and across muddy creeks. The footprints of the foxes were clear enough there in the snow, but even the briefest glimpses of the animals were rare. Fox hunting in my youth was a great way to develop

In northern latitudes, the willow ptarmigan is a common prey species of the fox. This Alaskan red is carrying away a spring bird.

strong legs and see a lot of lonely Appalachian back-country, but it never reduced the fox population.

One of my fellow hunters once claimed that a fox he was trailing jumped on the back of a sheep and "rode" it for a distance to throw a pack of dogs off its trail. I don't believe that story either, but I can clearly visualize the fox, sitting on a hilltop and laughing at the confused pack of dogs milling around below him.

Red foxes keep very busy even when they are not being pursued by hounds. For at least half of every year, family duties demand full time. Unless one or the other is killed, a male and female probably form a lifetime partnership. When wandering the Ohio hills in winter, I would as often find the tracks of two foxes wandering and hunting together as I would one. The pair would separate, hunt on roughly parallel tracks for some distance, and then rejoin. The later in winter, it seemed, the more often I would locate a pair, which shouldn't have been surprising. In the eastern United States, the breeding season occurs in February. The vixen will be in estrus for about three weeks. Copulation will take place many times during that period if wild red foxes behave similarly to captive animals.

During early March a female fox will spend more and more time searching for potential den sites. These may be her own previous dens, woodchuck holes, or natural cavities in the earth. I knew of one den inside a long-abandoned coal mine in eastern Kentucky. As the end of the 53-day gestation period approaches, the vixen "improves" as many as three or four likely den sites by cleaning them or adding fallen leaves or debris. That way, if one site is discovered after whelping, she can quickly move the kits to another site.

In the central United States six or seven kits are born during the last half of March. In Alaska three to five young are born a month or more later. As with many other species, red foxes reproduce at a rate to maintain their numbers wherever they live. Greater hunting pressure and predation in the heavily populated farm states, as opposed to the far North, probably explains the difference in litter sizes.

Kits average about 3½ ounces in weight and are blind at birth. Eyes open at between one and two weeks of age, but the kits do not resemble miniature adults until about five weeks old. That is when they take their first tentative look outside on warm days and begin playing in the sunshine. Some of the most pleasant times I have spent outdoors have been watching red-fox kits through a long-range spotting scope as they rough-housed at the den mouth. I advise anyone to make the most of such an opportunity because it rarely lasts longer than a day. During the night the mother may move the young to another den.

Obviously the weeks following emergence of kits from the den are the most precarious for them. No wild carnivore will pass up a chance to pounce on them outside. They are easy to dig or smoke out, or to poison or asphyxiate as a means of predator control. Fortunately nowadays, because bounties are no longer paid, digging out dens is falling out of favor.

Except for a period after kits are whelped, males and females refuse to den up. No matter how brutal the weather, both prefer to curl up in a ball somewhere outside, using the long tail as a blanket, covering nose and foot pads, but not eyes. If it is snowing, the snowfall just adds to the already efficient insulation of silky body hair. After whelping, however, the female remains in the den with the kits. The male serves as hunter and family provider, bringing back to the den anything remotely edible, from snakes and mice to farm poultry and beetles. This is a full-time occupation and, even if well done, survival of all the kits is far from guaranteed. The smallest, weakest ones may not get enough to eat and so will soon die.

In times past when American farming was a family endeavor rather than a corporate enterprise, and every farmyard contained chickens, a foraging red-fox male could be a menace as he harvested the springtime pullets for his young. Today most chickens are raised en masse in predator-proof enclosures where a virus is more likely to wipe them out than a fox is to catch a few. So *Vulpes fulva* must concentrate more on rodents than on domestic poultry, which should greatly please any farmer.

A healthy red fox is an almost indefatigable animal that spends a lot of its life on the move. When not actually chasing something or being chased, it travels at a steady trot for hours on end. Because its chest is narrow and its legs are located just off-center, a set

of fox tracks extending across the snow is a straight dotted line. When the fox is walking or trotting, the imprints of the hind feet fall right on top of those left by the front feet. Much can be learned about a fox's habits, its territory, its hunting strategy, and its prey species by following the animal across a winter landscape. If you are persistent and keep at it long enough, you may eventually get a glimpse of the fox moving a safe distance ahead of you. I am convinced that when a fox realizes it is being trailed, it will make a wide circle to get behind the pursuer for a better look.

Whereas foxes of eastern and middle America are extremely shy and wary of humans, I've found the foxes of the far northern wilderness to be much more confiding. At a fishing camp in northern Manitoba, one red fox would make an appearance almost every evening after anglers had returned from a day on the water. It would then stay in the vicinity, near enough to be photographed, until fish entrails or other scraps were tossed its way. A Cree Indian guide told me that this was a male feeding a family of pups denned in an esker not far away.

I guess I've seen more red foxes in Alaska than any-

This vixen (mother fox) watches as one of her kits toys with a mouse she has delivered. In summer here, her coat is shedding heavily.

A litter of four kits watched this vigorous "play" of adults from their den site just out of the picture.

where else. They hunt for Arctic ground squirrels along the road in Denali National Park, and a few of them learn to panhandle around the public campgrounds, showing little of the fear exhibited by foxes farther south. I have seen them along many of the salmon streams where the fish spawn on the Alaska Peninsula, feeding on scraps left by bears. On one occasion I saw a dark, cross fox frantically trying to capture a salmon in shallow water on its own. The animal raced back and forth in a frenzy, trying to pounce on a fish, but wasn't successful.

Once near a brown-bear-watching camp on Chenik Lagoon, Peggy and I had daily opportunities to enjoy the antics of a fox family denned near our tent. We could sit on a waterproof tarp less than 50 feet away and watch the playful actions of the kits while the parents were away hunting. One morning the parents returned, and the two of them also began to play-fight vigorously right in front of us. We've never seen any behavior exactly like it before or since. In Pennsylvania or Ohio the vixen would have moved her kits to another den far away rather than tolerate the presence of humans.

Fox researchers report that, like dogs, foxes sometimes bury any food they cannot eat right away. Like other canids they also have the need to establish urine scent posts to mark the limits of their home ranges, or at least of their recent presence. If after reaching maturity it established a territory of its own, and if it managed to outwit men and hounds, a red fox can continue to leave its scent posts on the land for 10 to 12 years.

My friend Murry Burnham described an interesting experience to me from a time when chickens still roosted in farmyard trees instead of in coops and mostly hustled their own food. One day when Murry was 10 years old, his Aunt Freda visited the Burnham home near Marble Falls, Texas, to report that a fox or foxes had virtually wiped out all of her flock. She was in tears, but she had come to the right place because Morton Burnham, Murry's father, was a legendary hunter in those parts and a game caller of unrivaled ability.

Morton and Murry picked up a couple of double-barreled shotguns and walked over to Freda's place. Murry hid behind a limestone rock while Morton sat behind another and started "lip-squeaking." He'd called only a few minutes when suddenly he was hit from behind and a red fox grabbed him by the wrist.

As one shotgun barrel went off, the elder Burnham instinctively tried to throw off the fox, which had a good grip on him. With a second hard jerk, the fox was thrown free and Morton shot at it with the other barrel as it ran away. He hit it, but didn't kill it. Now his wrist was bleeding profusely and both father and son hurried to town for a doctor to treat and bandage the wound. Wisely the doctor also began a series of rabies shots, because the fox's behavior was strictly that of a rabid animal.

Several days later, while still taking rabies shots, the elder Burnham returned to the same place and called in the fox again, this time killing it. He knew it was the same fox from the broken tail and the pellets he found in its rump. Meanwhile Morton Burnham's wrist wound festered and would not heal, eventually becoming so painful that he dug into the spot with his own knife blade. He found one of the fox's molar teeth embedded where it had broken off when the fox had bitten him. Murry admits he wouldn't believe the story if he hadn't been there.

Sylvatic rabies, or rabies in wildlife, is a matter of concern in America. The disease is not entirely

I was set up to photograph a brown bear feeding on the carcass of a sea lion washed ashore on Afognak Island, Alaska, But the first meat-eater to arrive was this red fox in the darker color phase, with the common name *cross fox.*

understood. Once domestic dogs were the main carriers of rabies, and the term "dog days" comes from this dangerous madness. But while immunization has more or less controlled rabies in dogs, the problem has shifted to wild skunks, bats, raccoons, and especially to foxes.

Says Dr. Samuel F. Scheidy, an authority on veterinary medicine: "Wild animals act in quite the opposite way from house dogs when they have rabies. A dog will sense that it is sick, draw away from strangers, wander away and try to hide. But when a normally man-shy animal such as a fox or skunk approaches you or a human habitation in a strange or friendly way, it's time to watch out." Although a person is far more likely to be struck by a car or mugged on a city street than to contract rabies, that word to the wise is best remembered.

**THE GRAY FOX.** Some years ago during a trip to Oklahoma, on a whim I bought a plastic varmint call and took it into dense brush country in the Panhandle to try it out. This seemed like the right kind of place to tempt a coyote or bobcat to betray its presence. I sat down where I could watch a long section of dry creek bed and began to blow the call without any practice or experience whatsoever. Not really expecting a response, I was startled when I immediately heard movement in the cover just beside me. An instant later, what I thought was a grayish house cat stood staring at me for a split second before vanishing. Then I realized that no cat ever had as thick and bushy a tail as that.

The "cat" was a gray fox, of course. Under such circumstances, it is easy to mistake a gray for a cat because this is the most catlike of the canines. In fact the gray fox has often been called the *cat* or *tree* fox, because *Urocyon cinereoargenteus* can and does climb trees. That long Latin name means silvery-gray-tailed dog.

In general appearance gray foxes are similar to their red North American cousins; in silhouette alone I'd be hard put to tell them apart. They have long narrowing snouts, luxuriant tails, and—compared with the red—relatively short legs. An adult gray stands about a foot tall at the shoulder and may weigh from 8 to 12 pounds, depending on a combination of geography and nutrition. The coat is grizzled salt-and-pepper with light underparts. The sides of the head, ears, flanks, and legs are a reddish or rust color. The tail has a black tip and black stripe along the top, and this streak may extend along the entire back. Although the coarser fur of the gray fox does not begin to match the silkiness and rich "feel" of a red fox's pelt—at least from a furrier's standpoint—I think that overall the gray is an equally striking and handsome animal.

The life cycle of the cat fox begins in a natal den that is usually located in the shelter of large boulders, in thick brush, or sometimes even above ground in the hollow of a large tree. The only den I have ever seen, in Florida, was well hidden beneath a rotting deadfall and overgrown with palmetto. No passerby could possibly have spotted it; this one had been detected by a black-and-tan coonhound with a very keen nose.

Three to five helpless pups are born in late winter or early spring after a gestation of about two months. The young are born with soft, short gray fur, and with eyes and ears closed. Apparently both parents share responsibility for care, feeding, and protection of the pups. When intruders appear near a den the adults bark and try to distract them. It may be more than coincidence that scattered observers report seeing gray foxes with broken legs; perhaps they are feigning injury to lure humans away from the den—very much the way some birds use the broken-wing trick.

Early life around the den is not unlike that of the red fox or most other carnivores. As they mature, the pups spend longer and longer periods playing outdoors and gaining strength. Young grays may also start hunting on their own earlier than other foxes, although mostly after grasshoppers and beetles. During a normal spring in gray-fox range, there is an abundance of these insects everywhere.

Gray foxes prosper on a diet even more varied than that of other canines. To determine their food habits and preferences in just one region, the Lower Sonoran Desert, biologist Frank Turkowski examined the contents of the digestive tracts of several hundred gray foxes taken in typical oak and pinyon-juniper habitat. The study revealed that although gray foxes are certainly meat eaters, they also like their vegetables, taking whatever is available locally and seasonally. Year-round, 65 percent of the desert foxes examined had small mammals in their diets. They also consumed birds, reptiles, anthropods (insects, scorpions, and spiders), and a variety of plants.

Among the mammalian foods, pocket mice as well as wood and kangaroo rats were more important than ground squirrels, white-footed mice, and rabbits. Ripening with the onset of fall, mesquite beans became increasingly important fare. Diet was most varied during the denning season because of the rapidly growing pups' increasing appetites, perhaps forcing the parents to be less selective. Thus more lizards, snakes, birds, and insects were gathered and eaten. Particularly interesting is the fact that 57 percent of the digestive tracts examined contained spiders, cen-

Aside from coat-color differences and its shorter legs, at first glance the gray
fox is hard to distinguish from the red.

Although the gray fox depends principally on small
mammals for food, it also consumes birds, reptiles,
insects, spiders, and a variety of plants.

The American gray fox is also called *cat fox*, ▷
owing to its stealthy movements and to
its ability to climb trees. (Steve Maslowski photo)

tipedes, or scorpions. Foxes consumed the latter complete with stingers and poison glands without suffering any apparent harm.

There is a good reason why foxes eat more anthropods and plants during certain times of the year. These may have been necessary to provide moisture during dry times when drinking water was not available in the Sonoran study area. But although prey species will vary according to region and habitat, the general types of foods in a gray fox's overall diet are essentially the same throughout its range. For example, juniper berries were most frequently eaten in Arizona, while in Texas persimmons and acorns made up a greater percentage of the vegetable food. In canine fashion, gray foxes everywhere bury surplus food and return later to eat it—unless a skunk, raven, or other scavenger has discovered it in the meantime.

The skull and especially the teeth of the gray fox (or any North American fox, for that matter) is evidence of its opportunistic life-style. Of moderate size, the canine teeth are not highly specialized as are those of the cougar and jaguar. They are instead good, all-purpose attack teeth. The molars in the rear are well suited for crushing tough vegetable matter.

Years ago a Georgia woodsman, who kept a succession of gray foxes around his cabin as semi-pets and for mouse control, told me about a fox's unique food-handling technique he had often observed. A gray fox can eat an egg of any size from a robin's to a wild turkey's with delicacy and precision. The fox does not simply bite into the egg or crush it. Instead the fox squats on hind legs, holds the egg between forepaws, cracks the top of the shell with front teeth, and then laps out the contents, wasting not a drop.

Shorter legs limit a gray fox's travel to a much smaller hunting or living range than, say, a coyote's or wolf's. Studies indicate that the gray fox has a home range of about two square miles. Although gray foxes can reach a top speed of about 25 miles per hour in a pinch, they are short-winded. Originally a forest species, they probably are now more abundant along the edge of cultivated lands than in totally wild places. No doubt they are attracted to this boundary area by the pocket gophers, ground squirrels, cottontail rabbits, and other rodents living there.

In addition to its ability and willingness to climb trees, the gray fox is unique in another respect: Its coarse bark (usually at night) is loud enough to suggest the voice of a much larger animal. I heard them when camping in my youth on catfishing expeditions and for many years wondered exactly what I'd heard. I got my answer years later, when I heard a pet gray bark. They also make whimpering sounds and growl.

Gray foxes are among the most interesting and use-

ful members of our wildlife community. They have never been as much at odds with farmers and ranchers as the other canids. What other creature can bark, climb trees, eat scorpions, suck eggs, and also wag its tail?

**THE SWIFT FOX.** During the mid-1980s, fur trappers in eastern Wyoming were surprised by a sudden change in events. All at once in eight counties they began to catch swift foxes again after not having seen any of the small, tawny canids for almost two decades. The species seemed to emerge from the dead elsewhere as well. But it was pretty easy to figure out what had happened.

The swift fox, *Vulpes velox,* is a native of the American Great Plains from the Texas Panhandle north to southern Alberta and Saskatchewan. Weighing only four to five pounds when full grown, the swift is only about half the weight the red or gray fox. This buff-yellow canid with a dark spot on each side of its snout is not as wary and suspicious as the other members of the dog family. It has always been more easily trapped. But the species managed to survive trapping. It wasn't until a far more lethal menace appeared that swift foxes began to vanish from much of their original range.

Compound 1080 was first mentioned near the end of Chapter 6. A poison designed to kill canids, particularly coyotes, the substance is also fatal to anything else that ingests it, including swift foxes. The poison was developed in the late 1940s and is a finely grained soluble white powder. Put a bit on your fingernail and it will slowly change character as it absorbs moisture from the air. First it will turn into a wet sugar-like mass, and then it will actually dissolve into the damp air. This solubility is one of 1080's most notable features.

Compound 1080 is odorless and, as far as anyone knows, tasteless. As such, it is presumed that there is no stimulus to any of the senses that might warn the animal eating a bait laced with 1080 that anything more than just the meal is present. Thousands of scavenging coyotes have dined on 1080-treated meat and died.

Compound 1080 does not produce a "pretty" death. It attacks the nervous system, accelerating the vital processes. Typically, a 1080-killed fox will end up on its side, all four legs "running" wildly as it dies. And death does not come quickly. Compound 1080 dooms the animal to several hours of agony before it finally succumbs. There is no antidote.

Commonly in range country, a bait animal (typically an old horse) would be killed and cut into about ten parts. While the meat was still warm, a water solution of 1080 was injected with a "brine gun" that features a long needle with holes along its sides. When the handle on the brine gun was pushed, the poison solution was forced into the warm meat. If done properly, and immediately after the death of the bait animal, the poison spread evenly through the entire meat portion.

Then the meat was allowed to sit for 10 to 20 hours to stiffen and seal so the poison would not escape. The bait was then taken to the desired bait station and secured to a tree or heavy stake.

If the bait was properly prepared, two or three bites were enough to seal the diner's fate. The poison would begin to take effect; the coyote or other creature would get nervous and excitable, and take off running. Rarely did a 1080-poisoned coyote or fox die at or near the bait station.

In my mind, the use of Compound 1080 was always a practice without conscience—a despicable practice. And it is to the shame of our federal and state governments that its manufacture and use were ever permitted. Not only did the poisoning toll on wild dogs continue for too many years, but cougars, black bears, bald and golden eagles, ravens, and even many songbirds were also killed unnecessarily. For the results of the poison did not end with the individual that first ingested it; any scavenger eating the original victim was also poisoned.

Under intense pressure from conservationists, President Nixon banned the use of 1080 in 1972. The swift fox is just one of the meat-eating species that has made a comeback since that time.

There is no absolute scientific evidence that removing the poison from the plains brought back the swift fox. Most stockmen certainly dismiss the idea. But Harry Harju, head of the Wyoming Game and Fish Department's biological services, believes there is a definite connection. He notes that the trapping harvest has more than doubled in the state during the past decade and that the foxes are appearing in former 1080 areas where they haven't been seen in a long time.

Dr. Major L. Boddicker, former fur-bearer biologist at Colorado State University, reminds us that the fortune and calamity of the swift fox have always been related to massive poisonings—to compound 1080 as well as earlier applications of strychnine and cyanide.

Perhaps because its pelt is of little value, and because it is not a game species, we have never bothered to learn much about one of the prairie's most beautiful and agile hunters. For example, we don't really know whether the swift fox is a nocturnal or a diurnal species. Not many outdoorsmen even realize that the swift fox is a separate species, rather than a description of any fox that is fast afoot. I have seen very few of them and always it was a blur of motion—a house-cat-size animal flashing away with tail straight out behind in a smooth-flowing, effortless stride until out of sight. Canadian friends who know them a lot better say that, when surprised, some swift foxes will hug the ground motionless to escape detection rather than try to run away.

Swift foxes may spend more time underground than other foxes, even when not raising a litter of pups. At least that is the consensus of trappers who have followed their tracks for long distances over snow-covered ground. A "home" den may be an abandoned badger or prairie-dog hole, but a swift fox may dig its own with both a front and back entrance for fast escape. Dens with up to eight entrances have been found in areas of soft soil.

Litters of four to seven pups are born from late February into early March. The young ones are raised and fed pretty much as are the other fox pups. They are nursed for about ten weeks and after that fed by both parents on small mammals, insects, gallinaceous birds, and small reptiles.

**THE KIT FOX.** Some mammalogists believe that the kit or desert fox, *Vulpes macrotis*, is simply a paler-colored race of the swift fox. Though it lives in a generally drier habitat, and perhaps has slightly larger ears, the kit is almost indistinguishable from the swift fox. Kit foxes are rarely seen, even by those who constantly prowl the deserts of the Southwest and Mexico. They remain in burrows during daylight hours, emerging to forage on creatures active at night. Judging from the fur, feathers, and bones usually found around kit fox burrows, the species apparently carries much of its prey home to eat near the safety of its retreat, rather than where it was caught.

In his book *Mammals of North America*, naturalist Victor Cahalane writes that the kit fox may be one of the fastest of mammals in a short chase. Such speed afoot is useful both in escaping from its own enemies— the golden eagle, coyote, and cougar—and in catching its prey. To escape pursuit it zigzags sharply at bewildering speed, running low with tail extended, then drops into any handy hole. Cahalane and many others have described how a kit fox can be running at full speed in one direction and suddenly dart in another without missing a step or losing velocity.

Despite its agility in eluding natural enemies, this elegant and evasive creature remains a comparatively unknown member of our desert fauna. No longer is it abundant anywhere, and especially in Mexico *zorra norteña* seems incapable of coping with guns and especially with the poisons still widely used there.

**THE ARCTIC FOX.** All along the northern fringe of North America, from beyond treeline to well out on the polar ice pack, roams the northernmost wild dog, a lonely animal covered most of the year with a pelage of the purest white. Mammalogists refer to it as *Alopex lagopus*, the Arctic, or white, fox. This fox is called *Psewkqoq* by the Alaskan Inuits and *Terigineak* by the eastern Canadian Eskimos. To both it has often been a major source of cash income.

An Arctic fox pelt hanging in a furrier's shop or at fur auction seems to have been peeled from a creature twice its size. That impression comes from its soft, long-haired coat that is as effective as camouflage in the winter wild as it is beautiful to see and feel. Actually this is a small canid, the body measuring only about 20 inches with a tail slightly more than half that length. Adults weigh from 8 to about 12 pounds, depending on age, sex, and time of year. People seeing white foxes in zoos for the first time tend to think of them as "cuddly" or puppylike, and perhaps less sinister or crafty looking than the other native foxes.

Arctic foxes are not always white, even in winter. There is a bluish phase from which the pelt is even more valuable than the usual pure white. All of the Arctic foxes on the two main Pribilof Islands—St. George and St. Paul—are the blue phase. In summer the short fur of white-phase animals is tawny in color, as compared to the dark brown, shorter summer hair of the blue phase. Spotted or dingy intermediates between blue and white phases occur throughout the range of the species. Thick hair grows on the feet of both in winter, concealing the pads and nails to keep the feet warm and permit easier travel over powder snows of the polar country.

The fur of the Arctic fox happens to be among the best natural insulators known to science. In fact, the fur is so effective that the outside temperature might plunge to −20 or −25°F before the metabolic rate of the animal increases. The species appears unaffected by temperatures as low as 60°F below zero.

Such adaption to environment affords the Arctic fox certain advantages. Rather than having to increase its caloric intake during severely cold winters, white foxes can continue to travel almost continually and hunt in those regions at the top of the earth where any kind of food is scarce.

At least while they are living on land, white foxes survive on birds (usually nestlings) and eggs, hares,

The fur on this Arctic white fox is among the best natural insulators known to science. The animal can survive long winters when the temperature remains far below zero for many weeks.

some berries, and scraps from the kills of larger carnivores, though mostly on lemmings. The latter are cyclic in number, building up every four years from very few to great abundance before dropping drastically again. Arctic-fox populations roughly follow these lemming cycles by a year or two.

The Arctic fox begins to breed when it is about ten months old. Its birthing den is usually on a dry, often south-facing slope in a rock pile or esker in the tundra with more than one entrance. The same dens may be used year after year for generations. The mating season, which Eskimos call the "barking" season, begins in February and lasts for a month or so. The mating animals are very vocal, and their squalling can be heard from far away on cold, clear nights. Some scientists who have lived among the foxes believe their barking has a ventriloquial quality, sounding farther away than it is. After a gestation of about 51 days, litters of from 1 to 14 pups are born, the number probably in ratio to the availability of food at the time. Though an astonishing 23 embryos were found in one female, mortality is high and only a few of the pups, no matter what the litter size, are raised to maturity.

Pups are born weighing 2 ounces each, with fuzzy brown fur and eyes closed. Perhaps even before the pups can see or move about, the parents may carry them to other dens nearer to better food supplies. Both adults feed the pups, which emerge from the den when they are about a month old and engage in typical playful fox-pup behavior. The family bonds weaken in the fall, and the pups gradually leave or are driven away to other territories. Almost certainly these young of the year are the foxes that make up most of the Eskimo trapper's annual harvest.

It would be difficult for white foxes to raise any pups at all if not for an important coincidence. Soon after the pups are whelped in spring, hordes of seabirds and shorebirds arrive from farther south to nest in the Arctic. Such seabirds as murres and guillemots, puffins auklets, gulls, and kittiwakes nest en masse in rookeries along the Arctic coast. Many of these crowded, noisy rookeries are among the most spectacular of all sights in North America. Some are also handy places to be raided both for nestlings and for eggs by parent foxes with hungry mouths to feed. Foxes also wait beneath the most inaccessible cliff sites for young birds to fall or be pushed over the edge, or to crash-land when attempting their first flights. Foxes take advantage of the bonanza to collect as many of the injured birds as they can, hiding those they cannot immediately eat or feed to their young, much as a squirrel hides nuts in the fall.

On several occasions I have also watched foxes prowling around and below the murre rookeries on

Left: Arctic-fox dens are located on south-facing esker slopes or beneath rock piles, and they usually have more than one entrance. Litter size may depend on availability of food in the region. Arctic foxes often depend on chicks that fall from remote seabird rookeries, such as those of the glaucous-winged gulls (top photo) or murres (above).

cliffs of Alaska's remote Pribilof Islands. But I wasn't prepared for what I saw one morning when I crawled on hands and knees, dragging camera and telephoto lens behind, to the edge of a precipice about 500 feet above the cold gray Bering Sea.

Hundreds of murres were nesting precariously side by side on thin ledges just beneath me. But the most surprising, thing I saw when peering over the edge was a fox staring back up at me from just a few feet below the rim of the cliff. It appeared almost guilty about being caught in the act of nest robbing. But I still cannot understand how it managed to climb down (or up) to that spot on the edge of eternity. The whole scene was too dizzying and the wind was blowing too hard for me to get a sharp enough picture of what I saw on that memorable day.

The Arctic fox is seldom shy. Many wanderers in

the Far North report that the animals sometimes follow and even bark at them. Other un-foxlike behavior includes several pairs occasionally denning in the same vicinity. One explanation is that suitable denning sites are not plentiful everywhere. A less likely possiblity is that white foxes are more polygamous than most observers now believe. It has always been accepted that these most northern foxes are at least seasonally monogamous and probably pair for as long as life for both continues.

For a portion of every year many white foxes become marine rather than land animals. For at least six months some may wander over the ice fields, their feet never touching solid ground. During his 1893–1896 trip over the polar pack, Norwegian explorer Fridtjof Nansen reported seeing fox tracks near 85 degrees north latitude—far, it would seem, beyond the point of finding anything whatever to eat. Most such vagabonds do not wander aimlessly, but instead one or a pair will follow a polar bear. When the bear kills a seal, it gorges until it is filled and then lies down to sleep it off. The fox then moves in to feed quickly on scraps. If a polar bear prospers, the fox—like the remoras that hitch rides on sharks—also does well.

One Arctic researcher explained how a roving polar bear in winter is really the equivalent of territory so far as the fox is concerned. Like a specific area of land, the bear "belongs" to one or possibly a pair of Arctic foxes. The resident fox will drive away other foxes that try to horn in on the action. A fox that does not follow its own bear in winter is like one that cannot find an unoccupied territory in July, and thus may not survive.

Late in winter breakup and melting ice may expose fish, dead whales, or seals that had been frozen in the ice pack. A fox can dine royally on such a find. As the Arctic winter blends into spring, a determined fox might also kill newly born baby seals on its own.

But the truth is that winter or summer, this species of fox is a more skillful scavenger than hunter. The carcass of any caribou fawn that was unable to keep up with its migrating herd will most likely be discovered first by an Arctic or a red fox. In June 1986, Peggy and I explored the misty wilderness around Af-
ognak Island, Alaska, with an old friend and bear hunting guide, Roy Randall. One day we came upon the carcass of a Steller's sea lion recently washed ashore on a beach at turbulent Cape Tolstoi. Mooring Roy's craft in a sheltered cove, we went ashore in an inflatable dinghy and about 50 yards from the carcass built a secure blind of the driftwood, old fish netting, and other flotsam we found. Roy guaranteed that the numerous huge brown bears of the Tolstoi area would quickly find the half-ton of meat. We were eventually able to photograph the bears tearing at the "bait," but the first mammal to arrive was a cross fox.

Arctic foxes have enemies other than starvation and trappers. The same polar bears on which they freeload in winter will dig out any den containing pups during their travels to summer dens on land. Larger, stronger red foxes have been known to kill Arctic foxes where their ranges overlap. Snowy owls, which may nest near fox dens, and probably gyrfalcons, will take pups that wander too far from the den area. Of course, no fox would pass up the chance to take the eggs or any owlets it finds unprotected in a snowy's nest.

Artic foxes once figured prominently in the fox-ranching business in Alaska. Most of the husbandry was done with blue-phase animals on offshore islands—around which water did not freeze and fences were not necessary—and there were plenty of fish to feed them. Foxes thus introduced to breed soon destroyed the nesting waterfowl populations on many islands. But that industry has all but vanished in the face of more efficient game ranching elsewhere.

Confiding as they usually are, white foxes are easy enough to trap for anyone with stamina enough to run a trapline in such remote, cold country. But more than one inexperienced hunter has become seriously ill when trying to eat the meat as emergency food. Cooking well may eliminate the possibilty of contracting trichinosis, but the high vitamin A content in the liver has a terribly toxic effect on both men and dogs.

*Alopex lagopus* is an animal few of us ever get to see in the wild; but it is certainly among the most attractive and durable creatures in all of North America.

# PART THREE
# THE BEARS

Peggy and I will probably always remember 1986 as our Year of the Bear because we spent so much time in bruin company. After photographing bears in the Yukon and Alaska from May through July, we came home to Wyoming to find a black bear frequenting our own premises. Although it defecated one night at the bottom of our back steps and left other sign all around, we didn't see that bruin near the house. But early one morning on a trip to pick up mail, I watched that bear digging in a meadow near our country post office. In late summer and early fall we encountered more bear activity in nearby Yellowstone National Park than we had in a long time.

Our bear summer began with a drive in our camper van up the Dempster Highway, Yukon Route 6. We traveled from the old gold-mining town of Dawson northward over 700 miles of frost-damaged, gravel wilderness road to Inuvik, near where the Mackenzie River empties into the Beaufort Sea.

From Dawson the road winds for 100 miles or so through the Ogilvie Mountains before dropping down onto level tundra. It was in the Ogilvies that we met the first and greatest number of the grizzly bears of the Dempster.

Our original reason for driving the Dempster was to see the extraordinary bird concentrations that occur in the area every May. Blizzards of waterfowl, shore birds, and passerines, as well as many raptors, come. Birds arrive from far to the south to nest and spend the summer in the Arctic. The road makes this spectacle accessible. But from the beginning grizzlies intruded on our bird watching. One liked to walk around our camp. We saw one pair play-fighting in a snow bank beside the road and another which, I believe, were stalking a beaver that was trying to strengthen its dam before the start of the spring runoff.

After the Dempster we drove to Denali National Park in Alaska where the grizzlies had just emerged from their winter's hibernation. They were everywhere. Twice we just missed seeing bears capture and eat young moose calves within sight of the park highway. After Denali we found more and more bruins to photograph on Kodiak and Afognak Islands before spending a week at the McNeil River State Game Sanctuary on the Alaska Peninsula. There is no

greater bear spectacle on earth than the one that takes place during the annual spawning run of dog salmon. Our Year of the Bear may also have been our most exciting year of wildlife photography ever.

Members of the family Ursidae, North American bears have comparatively low population densities and rates of reproduction. They are large and wide-ranging. All three species—black bear, *Ursus americanus;* grizzly or brown bear, *Ursus arctos;* and polar bear, *Thalarctos maritimus*—are dormant at least to some extent during winter. All breed in the spring, but due to delayed implantation, tiny cubs are born in midwinter. Because of their unpredictable nature, all bears pose enough real or imagined threats to humans to be exciting, endlessly fascinating creatures.

In fact, I divide the wilderness of North America into two separate categories: wilderness with bears and wilderness without them. There is never the same feeling of exhilaration when bears are absent as when I wander in an area where there's a chance to meet a bruin. How many people know that one and sometimes two species of bear lived everywhere in the original American wilderness when Columbus arrived?

The black bear is the most abundant and wide- ▷
spread of North American bruins. Populations
exist in 30 states and are strong over half of Canada.

# CHAPTER TEN
# THE BLACK BEAR

Almost from the day that European settlers arrived in the New World, Americans and black bears have been at odds. The real reason may be elusive because native Indians and bears had more or less evolved together in truce if not actually in harmony. Both were predators, omnivores really, who foraged for and ate the same things. It's true that killing a bear (like killing any other rival) was honored as an act of bravery, but mostly Indians coexisted with bears as the new European arrivals were never able to do. Perhaps the Pilgrims brought with them an old prejudice. In any case, black bear range and black bear numbers have declined ever since in direct ratio to white expansion and white settlement of North America. Frontier history of the United States is laced with tales of bear encounters and bear hunters, with heroes such as Davy Crockett.

## CROCKETT, BEAR HUNTER.

In 1811 the area around Knoxville, Tennessee, was beginning to fill up with people. So with a new wife, Crockett simply traveled farther afield in search of "free" land. He found it on a fork of the Elk River near the Alabama border and began the prodigious task of clearing virgin wilderness for farmland. But he soon realized that he was more adept at and stimulated by hunting than plowing and planting. He was an especially good bear hunter. Black bears were then so numerous in the forests and canebrakes of west Tennessee that Crockett was able to make a subsistence living by killing them.

The Crocketts ate the meat and used the fat for a variety of purposes. The prime skins were marketable back east. Crockett killed 115 bears in one year alone, or one bruin every 3½ days. No wonder he became a legend long before he was elected to Congress from Tennessee. He died at the Alamo while exploring Texas in search of still new wilderness to conquer.

One of the typical tavern tales, inspired by Tennessee sour mash, about woodsman Davy Crockett is worth repeating. So the story goes, one day Crockett was suddenly threatened by a large bear. But Crockett gave the critter such a fierce look, and the bear was so frightened, it froze in mid-charge. The legendary

*Excellent climbers, black bears sometimes spend long periods in trees. That may be to nap in summer breezes and to escape troublesome insects.*

woodsman had to build a fire to thaw the bear out. He then took it home where he taught it to smoke a pipe. Crockett called the animal Death Hug, and it is said the two vagabonded together thereafter.

Thanks to other legends, songs, and Disney movies, we can still visualize Crockett wrestling behemoth bears with knife and tomahawk. But the truth is that he was very prudent and methodical. In reality Crockett used a combination of iron traps, bear-meat bait, hound dogs, and an accurate muzzleloader as he cleared the land of black bears. Hand-to-hand hunting was dangerous and not Crockett's style. It is only in recent decades that we Americans have begun to develop a new philosophy about bears. Many of us no longer think the only good bear is a dead one. We want to be sure they survive on earth as long as we do.

## RANGE AND NUMBER.

Given suitable habitat, black bears are adaptable enough to survive without further assistance. That they have endured at all is a result of extreme shyness and above-average animal intelligence. Like most forest dwellers they are difficult if not impossible to see often, even by people who live in bear country. So a really accurate census of numbers is out of the question.

Writer-naturalist Ernest Thompson Seton estimated that two million black bears existed in the continental United States and Canada before settlement began. Many biologists today believe that figure was only a wild guess, and that the actual number may have been much higher—or lower. A consensus of bear scientists in 1985 set the black-bear population at about 200,000, or about one-tenth of Seton's estimate.

Today we have at least token populations in 30 of the 49 states where the species originally occurred. Black bears still roam in a few scattered, mostly mountainous parts of Mexico. Populations are still good in all Canadian provinces and territories.

The western United States account for most of the black bears' haunts left in America, because that is where humans are few and the largest tracts of wilderness or semi-wilderness stand undisturbed. Within the United States black-bear density is greatest in the Northwest. The number living in Washington State is estimated at between 25,000 and 30,000.

Not surprisingly, Washington is also the state where bears come into increasing conflict with people, or

rather with the timber industry. The bears are cited for great damage to newly planted trees on former forest areas that have been clear-cut. (I described this problem in detail in my earlier volume, *Erwin Bauer's Bear in Their World*, published by Outdoor Life Books). Indeed the Washington Forest Protective Association, an industrial group, has since 1975 employed a professional hunter as director of animal damage control. In his line of work, Ralph Flowers has killed as many as 90 bears per season, and more than 1,000 altogether. This huge total may be more than Davy Crockett's record.

On the bright side, Flowers has recently developed a special kind of bear pellet, which, when placed around a new forest plantation in five-gallon buckets and eaten by bears, is believed to keep bears from chewing bark. Test plots in the Olympic Peninsula show that bear damage dropped by as much as 87 to 100 percent in a few areas due to the pellet feeding. The state of Washington is also budgeting money to expand Flowers' feeding program and learn if it will work on a broad scale.

**PHYSICAL CHARACTERISTICS.** Some measure of black-bear size, at least in Pennsylvania, is indicated by the regulated hunter's harvest figures from 1985. Of 1,029 bears shot that fall, eight weighed over 500 pounds each, which means they probably average heavier than Rocky Mountain grizzlies. The largest of the eight was a spectacular 581 pounds.

Only about seven or eight of every ten black bears are black. The more easterly its home, the more likely the bear is to be black. Nearly all have a white or cream throat patch.

No other native mammal exists in so many color phases. Next to basic black, the most common pelage color is a chocolate brown. But black-bear fur also comes in all shades of brown including cinnamon, tan, and honey-blond. Although all cubs in a litter are likely to be the same color, that is not always the case. I once saw a dark-brown sow in Wyoming with triplets—one black, one brown, and one cinnamon.

Two striking subspecies of black bears live in restricted areas near the northern edge of black-bear range. A cream-colored Kermode bear roams on several misty coastal islands of British Columbia. Not an albino, this subspecies usually breeds true to its ghostlike pale hue. The other unique black bear is the silver-blue or smoke-colored glacier bear of Yakutat and the Glacier Bay area of Alaska. Of all black bears,

Black bears occur in many color phases from cream to pure black, the most common. A cinnamon-phase black bear is shown here.

this one may live in the most spectacular, most awesome of landscapes, in a region where great glaciers flow from far inland all the way to the Pacific Ocean.

Except for the claws, the hind footprint of a medium-size black bear is strikingly similar to a human print. It measures about seven or eight inches long and four inches wide. But often the heel of the bear's hind foot will not make an imprint unless the surface is soft earth or snow. When a black bear is running at its maximum speed of 25 to 30 miles per hour, the hind footprints will appear just in front of the forefeet. But we see a lot more black-bear tracks than we do the bears themselves.

In Alaska, western Canada, and portions of the northern Rockies, black bears share their range with grizzly bears. Because a black can be similar in color and, as noted earlier, nearly as large as many grizzlies, distinguishing between the two may not be easy. Identification is even more difficult on cloudy days or in dense woodlands, or when an animal is moving far away. Excitement can also make an inexperienced person mistake one for the other. A good many grizzlies in areas where grizzlies are protected are shot because of mistaken identity.

Certain field marks definitely separate the two species, however. The black bear does not have the noticeable shoulder hump of the grizzly when viewed from the side. Blacks also have longer, thinner snouts than grizzlies, which have wide, more concave faces. An experienced observer will also quickly note the more swaggering, swinging, pigeon-toed gait of the grizzly.

Small grizzlies can, with great effort, climb certain trees. But any bruin found far up in a tree or easily climbing one is sure to be a black bear. Its shorter, curved claws can grip most kinds of tree bark so well that the animal can almost run up a mature tree. Cubs climb as well as adults. All bears descend rear-first, clutching and clawing the trunk, though I once saw a 200-pound black bear jump out of a tree from a height of about ten feet and hit the ground running.

**FEEDING HABITS.** While many wild creatures eat only a few specific foods—such as the black-footed ferret, which feeds almost entirely on prairie dogs— black bears are not selective. "Its diet," one biologist joked, "is limited to everything that's edible." Bears especially relish meat, fresh or carrion, and will go to great extremes to obtain it. In September 1986 I had an excellent opportunity to see how much meat a black bear will eat when the supply is unlimited.

In the extreme northern part of Yellowstone National Park, the Gardiner River closely parallels the main highway through a dry canyon where many big-

This Yellowstone black bear found a mule deer doe that had been struck by a car and died in the Gardiner River. For more than five days the bear guarded the deer carcass and fed on it, never leaving the spot until all but bare bones was consumed. The bear covered the carcass with brush and then slept on top of it between meals.

game animals often concentrate. One day a mule-deer doe was struck by passing car, but managed to reach the middle of the Gardiner River before succumbing in the cold, swift, knee-deep current. Late in the afternoon a brown-phase black bear happened along, found the doe, and dragged it out onto the bank farthest from the road. It began to feed on the carcass immediately. I estimate the bear's weight at 300 to 350 pounds. The mule deer weighed about 130 to 140 pounds.

For five and a half days, as tourists and photographers watched from the opposite side of the river, the bear did not leave the spot. It fed intermittently on the venison, covering the carcass with brush and grass between meals, until the carcass was completely devoured. That bear ate approximately a third of its own weight, or about 20 pounds a day, in less than a week. Furthermore, the bruin did not tear fiercely into the

dens in hillsides, under deadfalls, or occasionally in natural caves or tree root systems. Only females with cubs have company over the winter.

Black-bear sows that were impregnated during the previous June have litters of one to four hairless cubs in January. The cubs weigh less than a pound apiece and live all winter on only the rich milk of the mother. They emerge at three to four months of age, coated with fur and weighing four to seven pounds. The young bears continue to nurse for another year, which serves as a natural means of birth control for the sow. Nursing also increases the cubs' chances for survival because they will benefit from the sow's protection and teaching until another litter is born, usually three years later.

Pennsylvania's 7,000 to 7,500 black bears are the most prolific, as well as the largest on average found anywhere. According to biologist Gary Alt, "In some places bears reach 7½ years old before they breed successfully. But in Pennsylvania, more than 40 percent of females begin to breed when 2½ years of age. Ample good food and a fairly mild climate contribute to the high reproductive rate."

**BEHAVIOR.** It is easy to read a lot of human characteristics, good and bad, into black bear behavior. One warm fall day in Yellowstone Park, I sat on a slope above Hellroaring Creek and watched the antics of a female with twin cubs about 100 feet below me. When I first spotted the family, the mother nursed the cubs and then played affectionately with them for a few minutes. Obviously all were having a happy time.

But eventually the sow tired of the roughhousing, firmly drove the cubs away, and curled up for a nap alone in the sunshine.

For awhile the cubs continued to play together, climbing and scuffling on the steep bank just behind the mother. Suddenly one lost its balance and came rolling down the bank, striking its mother hard in the small of the back. She was up on her feet in an instant, snarling and swatting. The last I saw of the trio, the female was striding away annoyed, at a pace so fast that the cubs could barely keep up with her. I could have witnessed the same scenario, I thought, in any park frequented by human mothers and their offspring.

Our appreciation for wild creatures, especially those considered dangerous, improves as we learn more about them. In the case of the black bear, the most knowledge may have been accumulated by Dr. Lynn Rogers of Minnesota, now the world's recognized expert on the species. He has been studying *Ursus americana* in the Superior National Forest since 1969.

In the public mind bears tend to be classified either

carcass, indicating great hunger. Instead it dined deliberately, if not delicately, snacking throughout the day. Of course, that scene might have been entirely different had a rival—especially a grizzly—arrived on the scene. In fact, I am surprised this lucky black bear was able to have the carcass all to itself. The bear drove one coyote away, and a few scavenging magpies managed a few tidbits, but otherwise the prize was not shared.

**BREEDING AND BIRTH.** Black bears are not really sociable beasts, but both males and females are promiscuous. Once springtime mating is complete, males go their separate ways and have nothing whatever to do with their families. Caring for and feeding cubs is solely a female role. In autumn, as food sources dwindle and as snow falls in the North, the bruins excavate solitary dens in deep forest. They make their

◁ A male and female black bear of the Arizona mountains engage in springtime mating play.

After mating, males and females will separate—the female needing to defend her cubs from all males.

as ferocious and ready to attack unsuspecting folk or gentle and cuddly as depicted in stuffed toys and cartoons. Rogers has learned that these most humanlike of North American mammals are neither of the above.

To begin, bears greatly prefer to avoid contact with people and go to great lengths to remain out of sight. People most often run into them when shortages of natural foods prompt bears to invade campgrounds and town garbage dumps for a meal. Indeed, black bears become easily addicted to human foods that are too readily available. The cubs of a female addicted to garbage will also become addicted. It is also around dumps and unclean campgrounds that bears are most likely to become testy and troublesome.

A black bear isn't built to be an efficient predator, as is a cougar or jaguar. The species isn't able to catch enough healthy animals to survive year-round on meat alone. Its formidable fangs and claws are used far more to tear apart rotting logs so it can lap up the insects inside. According to Rogers, despite the countless close encounters between black bears and people throughout the 20th century, there have been only 40 authenticated unprovoked attacks. This works out to one every two years. Eighteen have been fatal.

Peggy and I have met a good many black bears along Rocky Mountain trails without incident. The first bruin that stood up on hind legs at my approach many years ago was intimidating enough. But I soon realized that this stance was a way to see better through myopic eyes, rather than a sign of aggression. In Great Smoky

Black bear cubs emerge from the den at three to four months old, weighing four to seven pounds.

Mountains National Park I have seen individual bears make threatening gestures such as snorting and making short lunges toward hikers. But these threats were never followed up by further action. After almost two decades of continuous close exposure to black bears, Lynn Rogers has not been attacked.

During a 1986 interview at the U.S. Forest Service's North Central Experiment Station in St. Paul, Minnesota, Rogers described some of his experiences: "I've had to enter bear dens to tranquilize the animals, take blood samples and body temperature, affix them with radio collars, and even shoot photographs in the dimness with a flash. I have chased and treed bear cubs

The face of this young female black bear only faintly resembles that of gentle Smokey the Bear and Teddy Bear toys.

Black bears might be encountered on remote ▷ hiking trails in the scenic High Sierra backcountry of Yosemite National Park.

in order to get my hands on them to examine them. Once I lost my grip and fell on a bear below, striking its head with my camera and kicking it in the face. But it didn't retaliate."

Remember that Rogers has a great store of knowledge about bear behavior and psychology in his favor when handling bears. For instance, he knows that as forest dwellers with semi-retractile claws, black bears are adept tree climbers. Because the cubs can climb as well as the adults, bears evolved without the strong instinct to aggressively defend their young from enemies. By contrast, a grizzly mother, which commonly wanders treeless meadows accompanied by young that cannot climb well in any case, has been correctly described as a bundle of raw fury in defense of her cubs.

**HIBERNATION.** Medical as well as animal researchers have long been intrigued by bear hibernation. A hibernating bear in Canada, for example, doesn't emerge even once from its den in about six months to defecate or urinate, eat or drink. All the while the bear burns body fat, and its cholesterol level soars. Yet it has no problem with plaque buildup in its blood vessels or with gallstone formation. No

wonder scientists are beginning to study hibernation more intensely.

Researchers have found that hibernating black bears produce a special bile substance, ursodeoxycholic acid, that dissolves gallstones; the acid is now being tested for that purpose on humans. Also when in hibernation, a bear's kidneys essentially shut down, yet the animal is not poisoned by the accumulated urea the kidneys would normally filter from the blood. Instead, urea is reabsorbed from the bladder and broken down to release nitrogen molecules. The dormant bear somehow uses these to create new protein and preserve lean muscle tissue, even while eating nothing and losing up to a third of its body weight. If the processes behind these phenomena can ever be fully understood, they might in time be used to help human burn victims, those with kidney failure, and—one day—long-term space travelers.

**THE WORST ENEMY.** Other than man, wolves and grizzlies are an adult black bear's only natural enemies and this, of course, only where their ranges overlap. Unlike some other wild animals, bears are very hardy and rarely the victims of viruses and dis-

ease, even where they live in close proximity to civilization and might somehow pick up the organisms that sicken domestic stock. Cubs of course can fall prey to carnivorous animals such as wolverines, coyotes, and foxes.

But as sturdy and powerful as the animal may be, the black bear is exceedingly dependent on a fragile food supply. With a diet of at least 90 percent vegetable matter, it is most susceptible to the vagaries of weather, particularly to drought and its effects on nuts and berries. Rogers' research has clearly shown that when the usual food supply is low, the sexual maturity of bears may be delayed several years. Litters average smaller than usual, and intervals between litters are greater. In hard times, new cubs and even yearlings are less likely to survive. But the most obvious result of a poor food crop is that bears begin to travel. They leave home ranges and may wander great distances in search of sustenance. When that happens, odds increase that they will be shot as nuisance bears.

The events of 1985 are a good example of what happens when drought causes a failure in the available forage. Beginning that summer, thousands of black bears from Canada invaded Minnesota and Wisconsin, and as many as 2,000 of them were killed as nuisances. But in 1986 with food abundant again, the surviving bears retreated to familiar Canadian territories. By radio-tracking some of them, Lynn Rogers learned that some bears had traveled as far as 126 miles from home and then returned when the food supply was again up to its usual level.

There is a puzzling aftermath to the 1985 black-bear invasion. Just a year later in Minnesota, two bears bigger than any that had been taken in the area were shot by sportsmen. That September a young hunter shot one near Mizpah that weighed 687 pounds live weight. Naturally he thought he had a new state, if not world, record. But a week later Dana Haagenson of St. Louis Park shot a Bunyan-size bruin along the famous Gunflint Trail. He field-dressed it on the spot, and friends helped him load the animal into the back of his pickup.

Unable to lift and weigh the bear on any regular scales, Haagenson took it to a mechanical hoist at the Department of Natural Resources in Grand Marais. There the huge bear officially weighed 632 pounds. It would have weighed about 758 pounds when alive since about 17 percent of a bear's body weight is lost when the entrails are removed. One state official said that this Minnesota record for weight was probably due to the animal's habituating a garbage dump in the vicinity of Flour Lake.

During that 1985 migration southward from Canada, many nuisance bears were live-trapped in culvert traps and released elsewhere. But trapping studies in Minnesota only confirmed what other wildlife managers had already discovered: black bears had to be moved at least 40 miles away just to assure that half of them would not return to the spot where they were trapped. In the Rocky Mountain states, problem bears that have been removed by helicopter as far as 50 or 60 miles needed only a few days to return.

**THE FUTURE.** What is the black bear's place in America? Why do we need them at all? That's a question too often asked.

*Ursus americanus* is an elusive game animal that furnishes meat and exercise and trophies for many sportsmen every year and, as such, is economically important. The black bear is also crucially important for other reasons. Wherever it survives, it is a constantly traveling dispenser of wild fruit and grass seeds, which emerge from the animal's digestive tract intact and primed for germination. Many natural ecosystems continue to prosper because the black bear is a part of them.

As I write this, plans are being considered to use black bears in what may seem an unorthodox project to save endangered grizzlies. At one time grizzlies were fairly abundant in the Cabinet Mountains of northwestern Montana. But a summer-long search in 1986 turned up few traces of the species. Investigators managed to trap only one alive, and a total of only three were taken during the preceding four summers. One of those was a 12-year-old male, one a 27-year-old male (later killed by archers), and another a 31-year-old female trapped twice. Montana biologist Wayne Kasworm believed she was the oldest grizzly ever recorded in the wild and guessed she couldn't survive much longer. The sow weighed only 190 pounds when first trapped in 1983 at 27 years, and an emaciated 150 pounds in 1986.

Kasworm and his aides found sign of only one grizzly cub in the Cabinet Mountains. So he proposed using black bear mothers as foster parents for grizzly cubs obtained elsewhere, in an effort to restore *Ursus arctos horribilis* to its former range. An even more radical proposal was to replace a black bear's own cubs in their winter den with healthy grizzly bear cubs born at the same time in zoos.

The idea certainly seems worth trying, perhaps over and over. Some of these schemes might work.

A mother grizzly studies her surroundings while her 1½-year-old cub grazes on the new grass of early summer. Throughout its life, a grizzly will probably consume more grass than any other food.

CHAPTER ELEVEN
# THE GRIZZLY BEAR

Late on October 7, 1986, officials in Yellowstone National Park learned that a photographer had been missing for three days. William Tesinsky of Great Falls, Montana, was known to have been in the park. The next morning authorities began a search, found Tesinsky's empty car parked on the Grand Loop Road, and at the same time picked up signals from the radio collar on a female grizzly bear known as No. 59. Worried over the coincidence of the missing man and the proximity of No. 59, the rangers zeroed in on the signal. They did not have far to go.

Less than a half mile from the car, according to chief ranger Don Sholly, the men came upon a grizzly standing over a body. "When it saw us," he said, "the bear began dragging it away. What we saw was the feet and shoes." Not knowing whether the person was dead or alive, the rangers went for a rifle and killed the animal, but it had already eaten the upper torso of the man.

Near the scene of the incident rangers found an elk bugle, as well as Tesinsky's camera still attached to a tripod. Twenty-one frames of the 36-exposure film roll inside had been exposed. Since there were almost certainly no eyewitnesses to whatever had happened, Sholly had the roll of film rushed to California for processing. He figured that for once in such a case, the film might reveal exactly what took place. Was Tesinsky following and photographing the bear, eventually pressing too close? Was it an unprovoked attack? Or did the photographer die of other causes, a heart attack perhaps, and was he found dead by the bear?

Tesinsky's death came about three weeks after Yellowstone Superintendent Robert Barbee had issued a warning about increased bear activity. Barbee noted in a press release that a poor crop of whitebark pine nuts, on which Yellowstone bears depend every fall, was driving the animals to lower elevations in search of alternate foods. Barbee urged all vistors to be extra cautious.

Bear No. 59 was fairly well known to park authorities and even to a few frequent park visitors. Several times Peggy and I saw the medium-size female together with her twin cubs, usually in the meadows of the Yellowstone River. Park officers regarded No.

59 as a "neutral" bear; she had exhibited no fear of people and did not avoid them, yet had never confronted anyone.

The bear had been trapped and released four times, first in 1980. The last time she had been captured was on September 4, 1986, together with her two cubs of the year at the Canyon Campground. She had learned to manipulate "bear-proof" garbage cans and was teaching her cubs the bad habit. She was moved to the eastern wilderness of the park, but three weeks later she was back again, this time without her offspring. Park authorities wondered about that.

During the late spring the bear had been observed chasing elk calves born at the time, which is perfectly natural behavior. But as the summer wore on, she became increasingly indifferent to the presence of humans. She was often seen beside park roads, grazing as if in a remote wilderness. No. 59 provided a wonderful opportunity for literally thousands of passing tourists to see a rare grizzly. She also seemed to frequent the developed area around Canyon more and more.

A few days after No. 59 was shot, the developed film returned from California, and rangers eagerly examined the photos for clues. But the roll contained only two pictures taken in Yellowstone Park, and both were of bison and a final shot of a dark surface that was out of focus. An attacking bear? Months later park officials issued a statement saying that Tesinsky had indeed been trying to photograph the grizzly.

As reported in *Audubon* magazine, the investigating board noted the following as among factors indicating that Tesinsky had approached No. 59 too close and thus provoked an attack, rather than that No. 59 had deliberately stalked him: "Tesinsky's camera was covered with dried blood and smeared with mud and grass. The camera was attached to the tripod. The legs of the tripod were fully extended, though one of the legs was bent and twisted. 'The bend,' said the board, 'appeared to be one that would occur when the tripod was struck while set up with the legs firmly braced'." In addition, Tesinsky was using a 80–200mm zoom lens, which is too short for photographing grizzlies. That is, to make a grizzly fill the film frame, a photographer using a 80–200mm lens would have to approach much too close for safety.

**SHEEP-KILLING SOW KILLED.** The previous incident was only one involving grizzlies in the northern

The grizzly, along with the polar bear, is the largest meat-eater that walks the earth. It is really more an omnivore than a carnivore, however.

Rocky Mountains during the fall of 1986 and spring of 1987. More than one expert on the species has stated that the Yellowstone ecosystem cannot lose more than one or two reproducing sows a year without dooming the future of the bears in the region. Indeed, there may be as few as 30 sows of cub-bearing age in the entire region.

One dark night on her ranch in the Yellowstone ecosystem, a woman heard her dogs barking. She also heard strange noises coming from the sheep pen near her house. She grabbed a flashlight and rifle and hurried to the spot, where she found bears standing over the carcass of a sheep. She started firing from about 25 yards away. When the rifle was empty a second grizzly female, No. 76, and one of her two cubs lay dead. Another Yellowstone sow was lost.

When first interviewed by state and federal wildlife officials, the woman said that she did not know the bears were grizzlies when she shot them. She intimated that she might not have shot had she known they were endangered grizzlies rather than black bears. But later she insisted that she was shooting in defense of her own life, rather than in defense of her sheep. On that plea the killing was legally justified by federal officials, and the case was closed both by federal and state authorities.

While Montana state law allows people to shoot grizzly bears when their property is in danger, U.S. federal law allows the shooting of grizzlies only in self-defense. It would certainly be hard for me to justify killing an irreplaceable sow and and cub for the sin of dining on sheep. Representatives of several national conservation organizations pointed out that the failure to prosecute in this case could be sending a signal that anyone can get away with killing grizzlies by claiming self-defense.

**BEAR 104.** In 1983, still another young female grizzly was trapped near the east gate of Yellowstone Park and marked No. 104. The bear has lived intermittently in that vicinity ever since. Throughout 1986 she was so visible in that area with her two new cubs— and at times attracted such a following around Pahaska campground—that regulars there dubbed her Arnie after Arnold Palmer, the golfer who also attracted a crowd.

Wyoming Game and Fish Department bear biologist Larry Roop is among those who kept a professional eye on 104. He noted that a few people continually hounded that sow, sometimes approaching as near as

This grizzly lolls almost like a human in a warm spring pool of Yellowstone Park.

20 feet and then yelling to get her to stand up. Roop thus surmised that she was relatively unafraid and tolerant of people—a mixed blessing, possibly an ominous one.

"She is also a very aggressive bear when she wants to be," Roop revealed. "I've seen her in some vicious fights and once saw her chase off a big menacing male bear that came around. She's a little terror."

Like No. 59, Arnie was regarded as a safe or neutral bear throughout the summer of 1986. But Roop, for one, was worried about the bruin's seeming lack of concern for the people so frequently around her. "One of these times she is going to get a little ticked off and someone is going to get hurt."

Roop also pointed out that of the seven people killed in Glacier Park during the last 20 years, most were attacked by sub-adult bears that as cubs were classified as neutral. As do other bear biologists, Roop sees some neutral bears as time bombs.

**ANOTHER PHOTOGRAPHER KILLED.** Another fatal conflict between bear and man occurred in Glacier National Park, Montana on April 25, 1987. In this case too, the man was a Montana wildlife photographer. He was Charles Gibbs, 40, of Libby.

According to the widow, Glenda, the couple had spotted a grizzly sow and cubs on Elk Mountain while hiking on that Saturday. Mr. Gibbs set out alone to photograph the bear and failed to rendezvous later with his wife as planned.

Evidence found at the scene of the incident indicated that Gibbs had tried to climb a tree, but had been pulled down. His body showed numerous bites and scratches on arms, legs, and head. Gibbs's body was found about 125 yards from where he was initially attacked.

There were no plans to destroy the mother grizzly who apparently acted to protect her three yearling cubs.

Again, it is my view that many aspiring wildlife photographers, having seen full-frame published photos of the bruins, fail to realize that those photos were taken with a very long telephoto lens. They attempt to duplicate them with a lens that is too short to allow photography from a safe distance.

In addition, I feel that the greatest danger exists when one person, or at most two, approaches a so-called neutral bear, acting as a competing bear might. Just such a situation may have occurred when No. 59

A female brown bear and her cubs carefully approach a salmon stream where larger bears are already gathered. The more bears in one spot, the more potential hazards for the cubs.

turned on the photographer in Yellowstone Park. It seems unlikely that a neutral bear would attack a crowd of followers, but it may feel confident enough to turn on one or two intruders. The same logic explains why an angry man will fight one tormentor with vigor but decide not to take on a gang of them. There are few records of bears ever attacking more than two people at once.

**RANGE AND NUMBER.** The North American grizzly is divided into two somewhat different subspecies. One, *Ursus arctos horribilis*, is the bear of the Rocky Mountains, interior western Canada, and Alaska. Before the New World was settled, grizzlies lived at least in small numbers over roughly the western two-thirds of the continent, ranging southward into Chihuahua in northern Mexico. But only north of the Canadian border is the species more than a small fraction of what it was in the past. The population in Montana, Wyoming, Idaho, and perhaps a corner of Washington may be as low as 400 and is unlikely to be greater than the most optimistic calculation of 1,200 animals. Nearly all survivors south of Canada are concentrated in Yellowstone and Glacier National Parks and the wilderness areas that immediately surround them.

The other subspecies, *Ursus arctos middendorfi*, is known as the brown bear, Alaska brown bear, fish bear, coastal or Peninsula bear, or—probably most often—the Kodiak bear. It is a larger grizzly race of coastal British Columbia and Alaska, differing from inland grizzlies mostly in its greater body size and growth rate. The more abundant available prey and longer growing season of Kodiak-bear habitat may explain the evolutionary difference.

Until 1959, after the species had been eliminated from most of its range, almost all our information about Rocky Mountain grizzlies came from cattlemen and hunters who had studied the animals only briefly over gunsights. But then in 1959 brothers John and Frank Craighead began an ambitious study in Yellowstone Park that lasted until 1966. During that time the two biologists and their assistants handled most of the 391 grizzlies then living in the park a total of 600 times. They stalked and followed, live-trapped and radio-collared, observed and examined in hand many of the bruins from birth until death. Today most of what we know about these bears of the northern Rockies is a result of the Craighead studies, which

Frank describes in detail in his excellent book *Track of the Grizzly.*

**THE PHYSICAL SPECIMEN.** In a way, it is easy to understand human awe and fear of grizzlies because they are one of Nature's masterpieces. The grizzly moves with a speed, arrogance, and even grace that is astounding in such a large animal. Along with the polar bear the grizzly is the largest land carnivore clinging to existence on earth. Despite its bulk, a wild grizzly bear can beat the world's fastest human in a 100-yard dash by 30 yards. A grizzly standing 150 yards distant would need only about ten seconds to be on top of you—if that is what it wanted. Fortunately most would rather stay 150 yards away.

During the past decade or so a dozen people have been killed by grizzlies, and predictably the outcry has been great. But recently sociologist A. B. Goldman estimated that during the same period over 100,000 people perished in automobile mishaps for every bear fatality; yet we still produce, purchase, and use fast motorcars. The ratio of drug- and alcohol-related deaths to bear fatalities could be even greater, yet people continue to drink and use drugs. Goldman also estimated that jealous husbands were from 15,000 to 20,000 times as lethal as bears in the United States. Even honeybees and ticks take a greater toll.

Grizzly bears are extremely well equipped to sense the approach or presence of trouble. A bruin hears very well and, when the wind is favorable, a grizzly can sniff out carrion, another bear, or a human from a great distance. Although its vision may not be quite the equivalent of 20/20 human eyesight, it is a lot better than many believe.

The habit of a grizzly to hold its head high as if straining to see, or not look up at all, is no sure sign that the animal is having trouble seeing or is unaware. Placement of the bear's eyes in its head is evidence to the contrary.

Consider that most small mammals as well as ungulates—deer, wild sheep, antelope—have eyes well out on the sides of their heads to see well all around. Conversely, all of the meat-eaters described in this book have eyes set relatively close together and facing forward. The result is binocular depth perception and accurate location and focus on prey. Though over most of their range grizzlies are only part-time predators, their eyesight is still acute enough for hunting and watching for danger.

◁ A grizzly pauses in digging for ground squirrels along the Denali Highway to stare into a telephoto lens aimed from our car. The next instant it was digging furiously again.

Next page: After having stood in the fast water of an Alaskan stream, watching the surge of salmon fresh from the ocean, this brown bear emerges with its catch, squirting eggs.

**HABITAT AND FEEDING HABITS.** Inland grizzlies prefer a habitat that is a mixture of forests and open areas. In Yellowstone and Glacier National Parks they might be found at any or all elevations, although they are more likely to be higher than lower during midsummer. Mostly the bears follow the natural food crops as these become available to them. Find a place, for instance, where blueberries are abundant and ripe, and you will also find grizzly bears.

Standing squarely atop the natural food chain, grizzly bears are never picky eaters. The animals of Yellowstone munch willingly on anything from whitebark pine nuts and huckleberries to bitterroot and grass, ground squirrels and field mice, elk and carrion, and the cutthroat trout that spawn each spring in the shallow feeder creeks of Yellowstone Lake. They also have the unfortunate habit of pillaging human foods that are too readily available.

In Yellowstone Park, for example—from when the first tourist accommodations were built in the mid-1800s through the end of the Craighead study in 1966—grizzlies lived at least in part on human leftovers. Indeed, they became accustomed to visiting garbage dumps just as their cousins, the Alaskan brown bears, regularly visit salmon spawning rivers. Recently in the town of West Yellowstone on the park's border, a bear found a winter's cache of dog food in a tool shed behind a home on the edge of town. The bruin ate or carried away almost all of the high protein pellets before being discovered.

Exactly what is the grizzly's impact as a predator? Over most of the species' existing range, its effect is probably minimal. Where grizzly and black-bear ranges overlap, the two species coexist mostly by avoidance, keeping to different habitat types in the same ecosystem. Blacks tend to prefer relatively forested places. But I have no doubt that if given the opportunity adults of either species wouldn't pass up the chance to snatch the other's cub. Meat is welcome in any form, especially if no risk is involved.

During the early summer of 1986 I photographed two sows with their cubs feeding on moose calves they had just killed in Denali National Park. Peggy and I also photographed a thin, wounded adult cow moose that, as nearly as we could tell, seemed to have escaped from a grizzly attack. It is also well-known that grizzlies will try to capture both elk and bison calves during the early summer when dropped in great numbers. As might be suspected, snatching a bison calf from among the powerful mothers in a herd guarding them could

A mother grizzly nurses 2½-year-old cubs. Probably the longer the cubs remain with and learn from her, the better their chances of long-term survival.

be hazardous to a grizzly's health. But what is far from clear is whether the bruins actively hunt calves in spring, or simply take any they happen to find. Odds are that at least a few bears deliberately seek newborns each year. Cubs of females that are successful at this continue the practice themselves as adults.

Cow moose in Alaska's Denali National Park seem to have sensed that their offspring are safer from grizzly predation if the calves first see daylight in a region closer to human activity than the usual calving areas. But the bears now seem more and more willing to venture near these new calving spots. Almost certainly this boldness is learned behavior that will be passed on to future generations of bears.

Grizzly bears will try to dig out Arctic ground squirrels when coming across a colony of them. Less often they may try to excavate for marmots. On the Alaska Peninsula, I once saw where a brown bear had tried to reach a bald eagle's nest containing two young. But the cliff on which the nest was located was simply too precipitous for the bear to reach. Brown bears will also swim out to offshore islands to feed on the seabirds nesting in great rookeries there. Roy Randall, wilderness-lodge owner on Afognak Island (a close neighbor to Kodiak), tells of finding the remains of

sea otters on an offshore island. He said they had been eaten by a brown bear. Summed up, however, the toll of other wildlife taken by the bears is insignificant.

**DEVELOPMENT.** An individual bear's food preferences, indeed the entire pattern of its life, greatly depends on what the animal learns from its mother during its year or two as a cub. Biologists have noted that only man and some of the large apes rely more on maternal instruction for what they know and how they will react from that time onward. That grizzly bears tend to be individualists is a result of generations of individual, rather than group, teaching. The behavior of most other animals such as deer and elk is determined more by instinct, by their genetic inheritance, and all tend to behave much alike.

From one to four grizzly cubs (usually two) are born in midwinter in the mother's den, as are black-bear cubs. John and Frank Craighead found that though males outnumber female cubs by two to one, the sexes become about equal in number by three or four years of age. Half of all cubs do not survive to see their second summer. And male bears are a significant factor in these mortalities, even though it is widely believed no animal is more furious and determined in the de-

Here is a tender moment in the lives of a grizzly cub and mother as the young one rolls on its back, giving an invitation to play.

fense of her young than a grizzly mother. We have eyewitness accounts of females being killed by stronger males while trying to drive them from the young. Yet we know also that some mothers unaccountably leave cubs to their own devices when confronted by larger bears, or by humans. (I do not, however, suggest *ever* approaching a mother grizzly with cubs, for any reason.)

Grizzly cubs usually spend two years and sometimes three in "school." The teacher is very exacting, the lessons hard, and the discipline unbending. Older, more experienced mothers tend to make better teachers. Allowing for the difference in longevity, grizzly bear cubs go to school for about the same length of time as human children.

An excellent example of how well bear cubs retain what they have been taught comes from Doug Seus, a wild-animal trainer of Heber City, Utah. Seus has trained many species for their roles in motion pictures, television, and TV commercials, but he has specialized in bears. Among those were a pair of cubs he raised together, with Seus acting as mother. Eventually Seus could keep only one of the pair, and he had to relinquish the other to a zoo.

Several years passed before Seus visited the zoo and saw the other bruin again. But it recognized him and immediately went through its entire repertoire of moves on the trainer's command. It is thus reasonable to assume that a wild grizzly cub will remember what it learns while trailing and obeying its mother. That assumption is confirmed by biologist Larry Aumiller, who has spent many summers studying the brown bears of Alaska's McNeil River State Game Sanctuary at close range. He has noted that cubs use virtually the same salmon-fishing techniques used by their mother, no matter how efficient or clumsy.

Few wildlife families are more interesting to watch than a grizzly's or brown bear's, and Peggy and I have been lucky enough to do that many times. There seem to be three infallible rules in bringing up bear cubs: follow mother, obey mother, and imitate mother. Within that code the cubs are free to enjoy themselves as much as possible.

Having fun includes a lot of activities such as splashing in water and riding down lingering snow slopes on their bellies; but mostly it consists of play-fighting with one another or occasionally with mother. But the cubs had better quit bothering her immediately

Even when masses of salmon are moving upstream to spawn, a female may
pause to feed three hungry cubs on her rich, nutritious milk.

when she wants a few moments' peace. And they had better come the instant she calls with a stern "woof."

One late summer morning on the Alaska Peninsula, we watched a brown bear with twins traveling around the open shoreline of a wilderness lake. One cub pressed close behind the mother, but the other kept falling farther and farther behind. That seemed to be inviting danger, because there were a number of large males in the area. (Numerous studies in different parts of bear country reveal that male grizzlies may be the largest factor in cub mortality.) When the lagging cub failed to respond to woofing, the female suddenly ran back, grabbed it by the scruff of the neck, and shook it. Thereafter—at least as long as the family remained within range of our binoculars, both cubs traveled right behind their mother.

**TRAVEL.** The Craigheads learned that the animals are prodigious travelers, with males wandering farther and wider on the average than females. One Yellowstone boar, radio-tracked for years, had a summer range of about 168 square miles. Another was credited with a summer range of about 1,000 square miles. Overall, grizzly territories probably average smaller where more food is available and where the density of the bears is greater.

Despite the territorial system, the paths of bears often cross and re-cross during their incessant movement in search of food, usually without serious conflict. Frank Craighead told me that most bears in a regional population recognize one another by scent and other signs, if not exactly by sight. Every bear always tries to avoid all bears of superior size and rank, and they are usually successful. Trouble occurs when the bears somehow muddle upon one another.

**MATING.** There is a period in spring when not all grizzly bears seek to avoid each other. Boars are on a constant search for sows coming into estrus. They probably depend to a great extent on their extraordinary noses to home in on receptive females. Only when a male and a receptive female eventually meet on a remote mountainside, do emotions akin to human affection emerge between adults.

The mating season—when the female is physiologically and psychologically receptive to copulation—lasts two to three weeks. Breeding pairs may remain together throughout that period. But *Ursus arctos* is

These triplet brown-bear cubs feed on a salmon delivered by their mother.
Cubs do not willingly share food, and there is intense competition among them.

a promiscuous species, and the same scent of estrus that attracted one boar will no doubt carry to others and attract them, too. So the first male might well be replaced by a larger one, which might in turn be driven away by a still more dominant animal. Every male will try to breed with every female it can in the busy June schedule. The denser the population of grizzly bears, the more opportunity there is for promiscuity. There may also be greater opportunity for the strongest male bear to pass along its genes for the good of the race.

When rival males meet in mating season, some confrontations are resolved immediately by a turn of the head or other gesture that is clearly a sign that one bear acknowledges the dominance of the other. The lesser bear will then either leave to seek companionship elsewhere, or simply linger in the area at a safe distance as some observers have noted. But when two boars of nearly equal prowess meet, or if the aura of estrus is especially compelling, only a violent clash can settle the matter of which boar will do the mating.

It appears almost ritualistic for two males to stand on hind feet, each trying to tower over the other, to exchange bites and blows, and sometimes to lock jaws. If one is bowled over, he is wise to escape instantly to avoid being bitten deeply all over. Close examination of many of the large males live-trapped in Yellowstone revealed multiple scars and damage around the face and neck. I saw a male in Banff National Park, Alberta, just after breeding season there; the animal had a badly torn lip and an empty eye socket that was draining.

During the 1950s, Charlie Abou—an Indian hunting guide of northern British Columbia—described to me a grizzly-bear breeding incident he had witnessed in the Cassiar Mountains.

Apparently four males of fairly equal size had been lured to where a female had come into estrus. As he watched throughout one sunny afternoon, Abou said it was a scene of total turmoil. In a sort of free-for-all, the males rushed and threatened one another, yet none was driven away. Once one tried to mount the female, but the others attacked it, and more stand-up fighting and snarling followed. At dusk, as Abou was giving up his vigil to ride to his camp on Coldfish Lake, he saw a fifth male grizzly join the fray.

It was midmorning before Abou returned to the trysting site, but all he found was shredded vegetation

Two large grizzlies wrestle and roll in what is probably a ritual of the mating season. Otherwise, except for sows raising cubs and occasional meetings along salmon streams, grizzlies are loners.

These young males are play-fighting in the McNeil River, Alaska. The play is rough though and can leave the players bloodied.

This cub may have been a single birth, or it could ▷ have lost a sibling that had not followed mother, obeyed mother, or imitated mother.

and grizzly hair scattered over the ground. The bears had moved elsewhere, and unfortunately the guide did not have great enough interest to search further for them. It would be interesting to know how the mating matter was finally settled. How badly any were injured. Did the last bear to arrive prevail and, if so, was it because the bear was less fatigued than the others?

The actual mounting and breeding of a female takes place at intervals throughout the estrus period. Sometimes it is simply a hasty contact, seemingly almost an afterthought. But Frank Craighead reports that two of the marked bears in his Yellowstone study remained coupled for a full hour.

**THE GRIZZY'S FUTURE.** Early in the spring of 1986, a pair of young grizzlies just out of hibernation wandered from dens high in the northern Montana Rockies toward the attractive ranching community of Choteau. They knocked over several commercial beehives, lapping up the honey and larvae inside, and—according to conflicting reports—killed some sheep, ate dogfood, ripped the upholstery from a pickup truck, and destroyed apple trees. The bears were tranquilized by the Department of Fish, Wildlife and Parks and then helicoptered far into remote wil-

derness. But the bruins returned to the Choteau area again and were shot. What followed was a textbook example of what can happen when the forces of conservation, politics, and greed are mixed together and seasoned with old prejudices. At issue was no less than the fate of the grizzly bear on the lovely, windy, high plains of western Montana.

Once this bear with the long curved claws and the dished-in face was king, top of the food chain in the open Big Sky country of America. But since settlement of the vast region by ranchers, grizzlies have been pushed back into a last stronghold in the Glacier Park/Bob Marshall Wilderness Area ecosystem. According to Montana biologist Keith Auna, the bruins may be trying to move back into their former domain. He notes that early in the 20th century, forest fires stripped great areas in the adjacent mountains of old growth timber. Huckleberries and other second-growth bear foods took over. Today the Rockies have revegetated, and bears are instinctively looking for better conditions elsewhere—often too near the edge of human settlement.

The problem that angers many area residents is that they cannot lawfully shoot the grizzlies. The cost of protecting an endangered species sparks fiery debate in Choteau and surrounding areas. Some do not see the grizzly as being anything but a menace, no matter

In spring a grizzly just out of hibernation stalks
through Alaska's Denali National Park.

what. As a result, many bears are shot and buried as landowners take things into their own hands. The practice is called "The Triple 'S' Solution," for *shoot, shovel, and shut up.*

The dilemma has driven a wedge between a Pulitzer Prize-winning novelist and his son. The father is A. B. "Bud" Guthrie, Jr., author of *The Big Sky,* and outspoken friend of grizzly bears. His son, A. B. "Bert" Guthrie III, is a rancher who claims too many of his sheep have been recycled by those bears. He has also claimed that his father is a romanticist.

Replies the elder Guthrie: "It was the realists who gave us toxic wastes, tainted waters, poisoned air, and denuded country. Now they want to wipe out our bears. I wish they could see things in the correct proportions. The scuffle for money today imprisons their minds." At 85, Bud Guthrie is a clear-thinking man respected in his community. But he is pretty much a solitary voice on one side of the bear debate.

There are a few positive developments besides Bud Guthrie's appeal to local conscience. The Nature Conservancy has acquired a 12,000-acre area called Pine Butte Swamp Nature Preserve where the bears will have protection. It is the last place where the species may be seen in its original plains environment.

Another bright spot is the fact that Fish and Game officers are working with ranchers toward a better way to dispose of dead livestock. One reason bears may be attracted to the area is the enticing odor of sheep and cattle carcasses left out of human sight to rot.

Not long ago Montana schoolchildren were polled to select a state animal, and their vote went to the grizzly. But conservative legislators tried to block the choice, and one state senator tried to substitute the elk, claiming, "the mere presence of the grizzly is stopping growth and development in our state." But the children won out over that state senator, and Griz was officially adopted as the state animal.

We cannot ignore the fact that the fate of grizzly and brown bears is inexorably linked to the fate of our North American wilderness. Some creatures can survive without wilderness, but these giants simply cannot. If these great beasts are to endure, they must have extensive habitat in which to wander safely.

Female polar bears guide cubs for the first two or
three years of their perilous lives. These bruins
were photographed in late fall along the Hudson
Bay Coast near Churchill, Manitoba.

# CHAPTER TWELVE
# THE POLAR BEAR

It's summer and the family waits on land for fall freeze-up and a return to the ice. Cubs will usually stay with their mother into their second full winter, sometimes lingering with her into the third winter, by which time they are about two-thirds grown.

N̲o one could mistake the polar bear, *Thalarctos martimius*, for any other animal. Like other bruins it is large, muscular, and heavily built, with a long dense coat. But this bear is off-white or cream in color to match the pale landscape on which it lives most of the year.

Yet more than color and habitat distinguish the polar bear from other bears. Blacks and grizzlies, as we have seen, are omnivorous, able to survive on plants as well as meat. But the white is the most carnivorous of bears, and for this it is well equipped with fearsome teeth and claws. Several structural features distinguish it. It walks on the flat of its feet, and the long claws on all four feet cannot be retracted. The tail is so small as to go unnoticed. Its head, eyes, and ears are small. It has a long neck and, relative to the other bears, a "Roman" nose.

According to the rules of nature, all predators must be opportunists to survive. They must adjust to changing climates and to periodic scarcity and abundance of their prey. Today, they must also be able to adjust to the presence of man. For polar bears that means coping with the brutal climate on the top of the world, living from year to year whether the seals, which are its principal prey, are plentiful or not. It also means avoiding the humans who invade the Arctic in growing numbers in search of minerals and other resources of the land or sea. A polar bear's life is tough, and it's not getting any easier.

**THE PREDATOR.** In mid-March the Arctic sun rises and then hovers above the polar landscape like a cyclopean eye, beginning the short, early season of plenty for the polar bear. A large male traveling among floating islands of ice spots his fin-footed prey dozing on the ice in the distance in the yellow sunshine. The bear pauses to test the wind. Then it begins a stalk with all the stealth of a hungry tiger in a hot, humid woodland.

The boar's chest and forelegs hug the snow and his hind legs gently propel him, as he slides closer, inches at a time. Only the three black spots of eyes and nose betray the perfectly camouflaged white ghost. The seal lifts its head and looks around, its intuition working. The bear freezes in place. Lulled by the solar warmth, the seal nods again, fatally. Now the stalker explodes out of his crouch in the snow. Before the seal can dive back into open water, the bear is on top of it, crushing its skull and breaking its backbone.

Now the boar bear dines leisurely on the warm meal, probably remaining in the vicinity until all or almost all of the meat is gone. When he finally ambles away on flat blood-stained feet, an Arctic fox, which has been following all along, moves in to finish what scraps are left.

That same mid-March sun shines over another polar bear as well, this one a female just leaving her winter den on the mainland 200 miles due south of where the boar killed and ate the seal. The boar may even be the father of the twin cubs she leads out from underground into the sunlight for the first time. They are now about three months old. The sow leads the cubs, which Eskimos call *Ah tik toks,* down to the edge of the sea.

For a while the sow feeds on mollusks, crabs, shrimp, and perhaps a salad of seaweed. But then the strange, disturbing scent of man activates the female's protective instinct. Geologists have set up a camp nearby. So she plunges into the water and, with cubs clinging to her rump, swims out toward a distant ice floe. For the twins it is the beginning of a vagabond life that will not be interrupted even when the sun disappears in September.

**RANGE AND PHYSICAL FEATURES.** The great white bear, or ice bear, is at present fairly well protected all around the globe. An international treasure belonging to no one nation, the species inhabits the ice of the Northern Hemisphere, wandering the polar territories of Alaska, Canada, Greenland, Norway, and the Soviet Union. The bears swimming in zoo grottos from Chicago to Singapore are stellar attractions, but this is still an enigmatic creature relatively few people have ever seen in the wild. Although the polar bear has been studied much in recent years, our knowledge of the species remains incomplete.

No other large land mammal could tolerate the extremes of darkness and cold in the ice bear's polar environment. The animal can run up to 25 miles an hour across uneven ice; swim up to 300 miles, it has been claimed, over open frigid water; dig into ice and snow; and kill a quarter-ton bearded seal with ease. Only the Alaskan brown bear rivals the polar bear in average size. An adult has massive forepaws as large as dinner plates, but still the animal is as agile dragging a seal out of the water as is a mink carrying a small trout. The long-distance swimming is possible because of the thick and oily fur, which is extremely buoyant. The coat and the skin beneath it also serve as an efficient solar collector.

Most scientists who have observed polar bears agree that a polar bear's hearing is poor, that its eyesight is fair to good, but that its sense of smell is superb. The animal can scent seal blubber or a stranded dead whale from 20 miles away over the cold, uncontaminated polar air. A special eyelid that acts as a polarizer prevents the snowblindness that would otherwise be deadly on a dazzling-bright, all-ice landscape.

Males may occasionally exceed 1,000 pounds in weight, and mature females average about half that. So the animal requires a never-ending search for enough calories to survive.

It is a miracle of nature that a creature weighing only half a pound at birth in a winter den beneath the snow can ever reach such enormous size. Biologists live-captured and weighed one male near Churchill, Manitoba—probably very near where it was born— at just under 900 pounds. Perhaps the all-time record was a Canadian boar that was carefully weighed at 1,450 pounds by bear biologist Dr. Charles Jonkel in 1971. It is difficult to comprehend how many seals and other protein that monster must have eaten to reach his extraordinary size.

**MATING AND BIRTH.** A prized trophy of anyone who has ever killed a male polar bear is the baculum, or penis bone, which resembles an ivory hammer handle. That some boars are found to have fractured baculums has led to many native legends about violent mating rituals. But probably the only violence during breeding season takes place among rival males. Fighting may be intense and prolonged in any situation where two or more males happen to arrive in the vicinity of a female coming into estrus.

Polar-bear mating begins in April or May during the long, often sunny days when the ice pack is disintegrating. It may last until June when the sun never really dips below the horizon at the top of the earth. Among most mammal species in which the male is much larger than the female, males gather harems during the breeding season. But all North American bears are an exception to that. A male polar bear will mate with as many females as he can find during the

This bear waits along the rocky shore of Hudson Bay for freeze-up in the fall. Then the bears, as if on cue, disperse over the ice for a winter of feeding on seals. Pages 168–169: Come winter, a sow with cubs has a lot of mouths to feed.

season over his lifetime, and a female will copulate with many males. Of course, bear density, ice conditions, travel logistics, and perhaps simple luck of the draw all regulate the ice bears' promiscuous lifestyle. But one thing is certain: Male bears can detect the sweet smell of receptive females from at least as far away as they can perceive the presence of a bowhead whale washed up on a lonely shore—at least 20 miles.

Both bear biologists and Eskimo hunters have assured me that the breeding season is the ideal time to locate bears for study or for killing. One factor is the perpetual daylight in which the bears become hyperactive. Females coming into season seem to wander at a more frantic pace, leaving an easy-to-follow trail by pausing often to urinate. It is not unusual to find males converging from several different directions toward a sow making a trail. But that may prove to be wasted travel for all but the largest of the boars. Among the top candidates, the honor of mating goes to the winner of fierce combat.

When October arrives, and the long winter night begins to fall, pregnant females head toward land. Their destination is the northern edge of the continent or offshore islands that are traditional maternity denning areas. The other bears continue to wander the ice pack to hunt for seals, as they did all summer. Unlike blacks and grizzlies, polar bears do not hibernate.

A maternity den may be only a shallow depression in an Arctic pressure ridge or esker. But it is usually situated so as to be quickly covered by windblown snow, and it lies much deeper beneath the surface as the winter progresses. When she enters the den, the female should be able to survive the winter, often very fat and perhaps sleepy. Her body heat alone will maintain a temperature of up to 40°F inside the den if the insulating snow piles up deep outside, even though the Arctic temperature may be 20° to 60° below zero for several months. The greater the accumulated snow drift, the warmer the interior of the den.

By late December, eight months after their mother mated, one to three blind helpless cubs are born. Each is smaller than a fox squirrel or prairie dog. The mother may be only vaguely aware of the occurrence, if at all. Three or four more months will elapse before the young glimpse daylight for the first time.

From birth until March or April the cubs grow solely on the rich, thick, nourishing mother's milk. Biologists who have tasted it report it has a pleasant nutty

*These young males are sparring, testing each other in more or less friendly contests that prepare them for serious matches during the breeding season.*

flavor. When the family finally leaves the den, the cubs will have seen it for the last time. The female however, may return to a familiar place to have another litter in two, three, or even fours years' time.

Bear cubs must emerge from the den in early spring to survive, because that is also when seal pups are born. A mother bear must spend all her time, followed by her cubs, searching for snow and ice dens where seal young have been hidden by their parents. Baby seals are a vital part of a bear cub's diet during its early months. Finding the seals is easier for a more experienced female than for a first-time mother. Therefore, the young of an older sow have a better chance to survive than those of a new mother.

In time the hunting becomes a family activity. Most families remain together until sometime during their second winter of stalking out on the ice, when family ties weaken and the younger bears drift away. However, some family groups remain intact until their third winter together, by which time the cubs are about two-thirds grown. These offspring have the best chance for a long, healthy life.

**DENNING SITES.** Interestingly, polar bear denning sites are not evenly distributed over their circumpolar range. Instead they are concentrated in eight separate areas that are fairly well known to wildlife scientists. The nearest of those to the United States, as well as the most recently discovered, is one in northeastern Manitoba. There an estimated 50 denning sites used each winter lie between the Churchill and Nelson Rivers. Another smaller area is just south of the Nelson River. All these vital locations are rigorously guarded by the bears from human intrusion. Just possibly, other polar-bear denning areas remain to be discovered.

**BEARS AND MAN.** Polar bears are among those uncommon species that, when full-grown, have no natural enemies but man and others of their own kind. Many records exist of male bears killing smaller, weaker bears, especially cubs or females with cubs. Killer whales may take a swimming white bear now and then, but they cannot be considered significant predators of the bruins. And while starvation takes its toll of animals that manage to reach extreme old age, man ultimately decides whether the bears live or die.

Most of the time polar bears try to give human beings a wide berth. But occasionally, when they are very hungry, have become too familiar with man and his activities, or—rarely—for no reason we can discern, the bruins become man-killers. Edward Nelson, a naturalist, writes of a bear that approached two un-

A female and her cub wander over the rough
surface of the ice pack off Wainwright, Alaska.

armed Eskimos who were setting out seal nets near
Point Hope, Alaska. While trying to escape, the first
man tripped, fell to the ice, and held his breath. The
animal sniffed him from head to foot, and even pressed
its moist nose against the man's nose and lips. Suddenly the bear spotted the second Eskimo and lunged
at him while the first man sprang to his feet and ran
in terror toward his village. The man returned with
help, only to find the ice bear devouring his unfortunate partner.

Long ago when Eskimos lived primitive lives un-
colored by contact with European civilization, hunting
white bears was a ritual and a test, as well as a way
to obtain meat and the warmest possible blanket or
sleeping robe. During such hunts, the men would deliberately expose themselves to a bear. When the bruin
charged, their sled dogs darted in from all sides, snap-

ping at the bear's flanks, then dancing back from its
swinging paws. While the bear was beset and surrounded by the pack of snarling dogs, the hunters
closed in for the kill with short spears. As often as
not, the men were badly wounded for their courage—
or foolhardiness, depending on how you look at it.
By custom, the one who sighted the bear first received
the hide. The slayer received the meat, though usually
all in his village shared it in his honor. In later years
if bears were thus encountered, the coup de grace was
administered with a rifle fired from a safe distance,
but the meat was still the property of the one who
downed the bruin.

By the middle of the 20th century the ice bear's
greatest threat came from men hunting them with
rifles by boat or ski-equipped light aircraft. Various
governments have now regulated against all sport
hunting of white bears, though native peoples are still
permitted to hunt for subsistence. Yet the threat from
humans may be as great as or even greater than ever.

Wherever biologists have a chance to examine polar
bears in the wild, they invariably find the animals
contaminated with various levels of chlorinated hy-
drocarbons such as Endrin, Dieldrin, and DDT. There
is no mystery about how the toxins entered the bears'
bodies. The bruins eat the seals, which in turn have
eaten the polar fishes and other poisoned creatures at
the bottom of the food chain. When poison chemicals
such as herbicides and pesticides are used anywhere
in the world, residues trickle to the sea and are thus
distributed everywhere. There is no such thing as
limited use of poisons.

This is but one instance where the problems of a
wildlife species serve as a warning signal to all of us
that the earth's environment is being badly damaged.
I wish I could believe that the world's leaders, especially our own, were intelligent enough and sufficiently concerned about the implications for the future
to do something about it.

# PART FOUR
# THE MUSTELIDS

Unlikely as this may sound, mustelids, or musk-carriers, helped shape American history almost as much as the great bison herds or any of the other abundant wildlife early settlers found in North America. In the process, the animal was almost rendered extinct.

In 1741 Vitus Bering, the Russian explorer of Alaskan waters, became shipwrecked on a small island in the sea that bears his name. The half-starved crew of the *St. Peter* managed to survive a brutal winter by eating the carcasses of the "sea apes" they found thereabouts. They saved the skins of these strange animals.

Although Bering died, his crew built a smaller ship from the wreckage and somehow made their way to Kamchatka with about 900 sea-otter pelts. When the fur hunters of Siberia saw the cargo, their pulses must surely have raced. Here was fur worth vast fortunes if sold in China—where the mandarins wore lustrous fur gowns—or to royalty in St. Petersburg. During the next four decades, fleets of Russian fortune hunters plied North America's northwest coast killing hundreds of thousands of sea otters and fur seals, and being killed themselves by the treacherous seas and storms. Until Captain James Cook of England arrived in 1778 and traded cheap trinkets to the Indians for sea-otter pelts, the Russians had a monopoly.

News of the Cook expedition spurred an even wider, more frantic fur hunt. French vessels also headed for the American Northwest, and Spanish ships sailed up the California coast toward Alaska. Then a Boston skipper, Robert Gray, became the first of many Americans to join the hunt. Those Yankees rounded Cape Horn in clippers, traded for sea-otter pelts, and then traded the pelts in China for silk, porcelain, and tea. They then hurried home to complete the cycle of trade that would make them wealthy beyond their wildest dreams.

Through this commerce as much as anything else, the United States established itself as a power in international trading and in the Pacific. The Stars and Stripes became known around the world, and at about the same time civilization gained a foothold in Oregon country the sea otters, not surprisingly, seemed to suddenly disappear from the earth. Hunting ended only because almost no otters remained alive.

A few otters did survive in lonely bays protected by dangerous tidal rips and currents, where they could not be easily pursued. Thanks to total protection and redistribution, several thousand sea otters now exist in scattered herds from Monterrey Bay, California, northward to the Aleutians. But they do furnish a classic example of how men have been willing to expoit wild creatures around the world for quick riches and power. Some species have vanished because they were a nuisance. But many of the mustelid populations have been depleted simply because of their lustrous fur.

Members of the family Mustelidae—including weasels and mink, otters and badgers, skunks and martens, ferrets and wolverines, as well as fishers—are successful predators. They generally have long slender bodies, short legs, short rounded ears, and anal scent glands. In many species, males are distinctly larger than females. Some are specialists in their feeding habits, while others eat a variety of foods. The river otter is one of the specialists, almost totally committed to catching fish underwater. The generalist wolverine, by contrast, is a hunter or carrion eater that will also feed on berries or insect larvae.

Another common attribute of all mustelids is their ability to catch and kill prey much larger than they, themselves, in some cases several times their own size. Ounce for ounce, mustelids are probably the fiercest predators described in this book.

The wolverine is the largest of the land mustelids, ▷
or weasel family. Wolverines are the subjects of
astounding tales and legends, some of them true.

# CHAPTER THIRTEEN
# THE WOLVERINE

I spent the summer of 1952 exploring and fishing across the wilderness of northern Manitoba on assignment for *Outdoor Life*. One of my guides on that trip was Edgar Whitehead, a Cree Indian trapper. One night we slept in his log trapping camp on the shore of Reindeer Lake. Our sleeping bags were unrolled on air mattresses on one side of the cabin; on the other side was a rusted iron stove and all our supplies, including several dressed lake trout that we planned to smoke the next day.

During the night I was awakened from a sound sleep by the cabin door crashing open and then by the sound of an animal grunting and clawing its way through the makeshift kitchen. "Bear," I thought immediately. Whitehead fumbled for a flashlight, shined the beam on the intruder—a wolverine that was beginning to

eat one of the trout—and threw a boot at it. What followed remains indelible in my mind.

Although we shouted at the wolverine and threw all our shoes and several chunks of kindling at it, the animal hardly seemed aware of our presence. Instead, it just kept tearing at the trout and eating until it had finished off eight to ten pounds of trout. Then it scrambled out the door and into the darkness. For me, this was a very impressive introduction to a very determined animal.

**PHYSICAL CHARACTERISITCS.** Through the years, the wolverine, *Gulo luscus*, has impressed backwoods people enough to inspire great awe, fear, and admiration, as well as countless legends and tall tales. Every northwoods trapper has his own repertoire

◁ Rarely seen in the wild, wolverines are constantly traveling over the wilderness landscape, seemingly without pause.

Ferocious and determined as wolverines might be, legend has sometimes painted them larger than life. In fact, Yukon males average just 31 pounds.

of stories about his troubles with wolverines. A wolverine on anybody's trapline, according to many, is the worst thing that can possibly happen.

Certainly the wolverine is the largest and most formidable of the land mustelids, or weasel family. It is stout-bodied, bushy-tailed (unlike most others of its tribe), and bear-like in shape and movement. About the size of an average dog but with very short legs, the wolverine is basically dark brown, with golden-brown shoulders and a broad yellowish band that extends down each side and across its hips. Its crown is grayish, extremities black, and underparts dark brown, while the throat and chest are white. The overall impression in any light is of a very dark mammal that never really stops traveling.

Also called *glutton* and, in French *carcajou*, the wolverine's fur is extremely valuable, prized for parka trim and hoods. Contrary to popular belief, wolverine fur *does* collect frost as will any other fur. Its value lies in the fact that the coarse guard hair is so durable that it will not pull out or break off as will other furs when the accumulated frost is brushed away. Because wolverine fur is so much heavier than other available furs, however, it is not best suited for making an entire garment.

Of all mustelids, the wolverine is second only to the sea otter in size, though it is not nearly so heavy as is often claimed. During one study, A. M. Pearson and R. A. Rauch measured and weighed a total of 284 wolverines taken in the Yukon and Alaska. The adult Yukon males averaged just over 31 pounds and the Alaskan males just over 23. Yukon adult females averaged almost 21 pounds, while their Alaskan counterparts averaged only 15½. There is no explanation for the difference between the Alaskan and Yukon specimens. The largest Alaskan wolverine of all was a 38½ pounder. It measured 40 inches long, of which the tail was eight inches. Full grown males stand 15 to 17 inches at the shoulder, females an inch or two less.

I mentioned before that wolverines are the bane of trappers who run long traplines in winter, and that deserves some explanation. Once a gluttonous wolverine has located a trapline, the animal methodically follows it, eating anything it finds from pine squirrels to mink and pine martens, until it itself is somehow caught and killed. Yet catching one is seldom an easy matter. Even worse than robbing traplines is the animal's habit of fouling anything it cannot eat, including the steel traps.

Like all members of the mustelid family, which also includes skunks, the wolverine has scent glands inside its anus. The musk it secretes may be even more powerful than the skunk's. Perhaps fortunately, it neither

produces the amount secreted by the skunk, nor can the animal direct it as a spray. No other creature will touch any meat that has been tainted by the musk or urine of a wolverine. Nor will any animal go near a trap the wolverine has fouled. Decontaminating steel traps is difficult, unpleasant, and time-consuming; thus, it is no wonder trappers harbor little affection for wolverines.

Recently wildlife researchers in British Columbia made a fortunate discovery about the potency of wolverine urine. While they were experimentally live-trapping snowshoe hares, a wolverine appeared and raided the traps in half of the study area. The scientists quickly learned that hares would not go near any of the traps on which the wolverine had urinated or deposited its musk. But in the other half of the study area, hares continued to be caught.

As a result, the scientists realized that wolverine urine is an effective repellent for many animals, especially deer. This may lead to finding a chemical synthetic that can be sprayed to keep both deer and hares away from the seedlings in newly planted forests, or even from devouring the home gardener's produce.

**RANGE AND HABITAT.** Wolverines require vast wilderness areas where a diversity of prey and other, larger predators exist. The species still lives across the northern half of Canada, in addition to most of British Columbia, the Yukon, and Alaska. There is also a wolverine population in the Rocky Mountains of extreme northwestern Montana. Within this territory the animals occupy a wide range of wooded and mountainous areas, and the preservation of this undeveloped, roadless habitat is the key to their survival.

**FEEDING HABITS.** It is safe to say that a wolverine is able to kill anything it can catch of comparable size. That includes any kind of protein from snowshoe hares to ground squirrels, and from ptarmigan and other ground nesting birds to the calves of caribou and domestic lambs. A large part of its diet consists of carrion found at the kill sites of bears, wolves, or human hunters. Among the more common wolverine stories are those of wolverines driving wolves and bears away from their kills, and in some cases these may be true.

One grizzly bear-wolverine confrontation over a carcass was witnessed by a party of summertime hikers in Glacier National Park and reported by naturalist Asa Brooks.

It seems that a packer who had been supplying a park trail crew lost one of his pack mules when the animal lost its footing and fell over a cliff from the popular Highline Trail near Granite Park Chalet.

Not even brutal winter in the far North slows the
wolverine's unending search for food.

When Brooks led a hiking party along the trail above
the dead animal several days later, he spotted two
grizzlies and two wolverines squared off around the
mule carcass.

The larger grizzly fed on the mule while the second
bear paced uncertainly some distance behind. One of
the wolverines was also feeding on the mule while
the other waited nearby. Whether the feeding wolverine moved too near the bear or aroused it by some
other means is uncertain, but something suddenly
triggered the grizzly into attacking the animal with a
blow of its forepaw. The wolverine was knocked away,
its neck apparently broken, as hikers watched the
drama through binoculars.

According to Brooks, the second wolverine disappeared immediately into heavy timber; and the
larger grizzly continued to eat, the brief interruption
seemingly forgotten. But the smaller bear picked up
the limp wolverine, dragged it to a rock some distance
away, then stood up on hind legs and violently shook
it. The bear was probably transferring its frustration
at being denied a share of the mule meat.

On several occasions during the 1950s I visited the
late, respected naturalist Adolf Murie at his cabin in

what is now Denali National Park. He was studying
the wolves of the park and their relationship to the
Dall sheep, but during his daily observations he could
not help but meet many other mammals, including
the wolverine. From the tracks he found in the snow,
Murie discovered that wolverines would retreat or give
wolves a wide berth whenever their paths converged.
In his *Lives of Game Animals*, naturalist Ernest
Thompson Seton tells of a wolverine driving a wolf
from a deer it had just killed, and of wolverines chasing
coyotes and black bears from animal carcasses.

Wolverines readily climb trees. One winter Murie
found a moose carcass on which a wolverine was feeding, but more than once a wolf appeared to chase the
wolverine up a nearby tree. The fact that an animal
carcass is frozen solid by sub-zero temperatures does
not keep a wolverine (or a wolf) from gnawing away
at it. If the sun on a rare warm day begins to slightly
soften one part of the carcass, any feeding animal will
concentrate on that part.

We can only speculate on what would happen if
wolf and wolverine met far out in the open where
there was no retreat and no tree to climb. When attacked by sled dogs, a wolverine was seen to turn over

179

on its back and face the attackers with powerful ripping claws and jaws. That defensive position might also be enough to discourage a lone wolf. In another instance, in the Yukon, a wolverine was powerful enough to drag a chunk of moose weighing more than three times its own weight over a half mile uphill from a hunter's camp before it was run down and shot by a man on horseback.

**MATING AND BIRTH.** The consensus among those who have studied the species is that male wolverines are polygamous. That must require immense energy and determination since average home ranges across the continent are estimated at 150 to 250 square miles. The home territory of each wolverine in Glacier National Park, Montana, is calculated to be about 125,000 acres. Peak breeding season is believed to be May or early June, but I once saw a pair of normally antisocial adults running together during an early July pack trip into the Mt. Assiniboine area of southeastern British Columbia. While within my sight the two cavorted and behaved as a mating pair.

Young, with eyes and ears sealed for the first month, weigh 3½ ounces each and are born in natural cavities or in earthen dens excavated by the mother. Dens are not easy to find, but more than one trapper or backwoodsman has reported their being located near a large winter-killed carcass that furnished the mother with a handy food supply. Young wolverines nurse for eight to ten weeks, but meat is introduced into the diet well before nursing is cut off. By midsummer the young must be strong enough to follow the mother on hunting trips if they are to survive. Soon after the first snowfalls of autumn, each goes its own way.

**THE PREDATOR.** *Gulo luscus* comes about as close to perpetual motion as anything in the animal kingdom. The animals don't appear to be fast afoot. Except when feeding, however, they seem to move throughout the wilderness with back arched, in a steady, ground-covering lope of about six miles an hour. That is about twice as fast as a human can walk briskly on level ground. A wolverine maintains that pace uphill as well as down, on tundra as well as on firm footing. If there is snow on the ground, a wolverine leaves behind tracks that might be mistaken for a wolf's. To watch or to follow the tracks of this animal is to get the impression that it is driven always by an insatiable appetite for food or something else that lies just over the next ridge.

Some prey in fact may become exhausted by the

With claws that serve as well for climbing trees as for fighting and digging up rodents, the wolverine packs a multi-purpose tool arsenal.

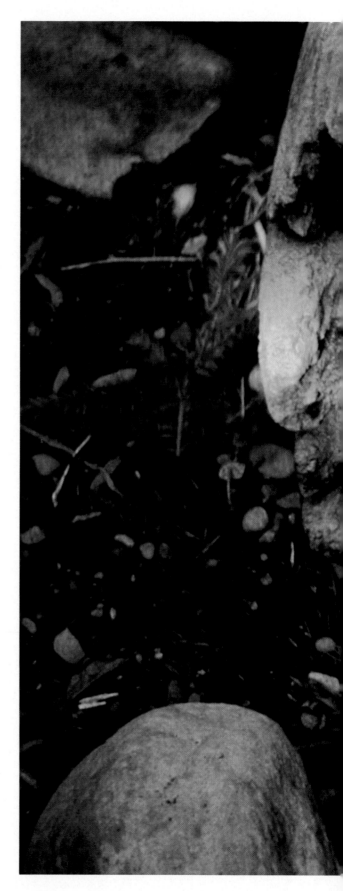

Perhaps the rarest sight of all is a wolverine resting, usually near the carcass
of a larger animal, which it will defend until eating the last of the meat.

wolverine's relentless pace, and then caught. Other prey, mice and ground squirrels, are taken by digging with the same long claws that permit the wolverine to climb trees so well. Very few wild creatures anywhere use claws for both purposes.

The food cache, a miniature cabin erected on tall stilts near every lonely backwoods cabin, is a fixture in the Far North. Trappers and other backwoodsmen have always needed these structures to store provisions out of reach of everything from mice to grizzly bears. But wolverines can get into the best and highest of them with little trouble, unless the poles are wrapped with smooth metal or stovepipe on which the animal cannot obtain a clawhold.

But wolverines often have their own caches to worry about. According to Adolf Murie, when a moose or caribou carcass is too large to be eaten entirely, rather than share, a wolverine will drag various parts of it away to be hidden for later consumption. Sometimes this works, but just as often it doesn't. No matter how meticulous or laborious the concealment, foxes especially are able to track and soon find a wolverine's treasure.

For the most part the glutton is a difficult mammal to see anywhere, at any time. Men who have spent their entire lives outdoors in wolverine country have seen very few, if any. Ernest Thompson Seton ad-

mitted seeing only two during his lifetime. And although Adolf Murie made tens of thousands of footprints while roaming across the excellent wolverine habitat of what is now Denali National Park, he confided that he did not see many individuals in any year's time. Murie believed they were more secretive than scarce, in an attempt to avoid humans. Were it not for the wolverine's penchant for breaking into backcountry cabins and caches for food, or for raiding traplines, the species would be even more a mystery than it already is.

A few young wolverines have been taken from dens and at least partially domesticated. In one case, a female made an affectionate and devoted pet. For two years she kept "her owner's" wilderness cabin area free of rodent pests, nor would she tolerate other animals of any size or species around the place. She was quite distrustful of strange humans, too. Then one day she simply vanished, leaving no clues as to why. She probably just loped out into life, in search of wolverine rather than human companionship.

The marten, or pine marten, is a long, lithe, quick
mustelid as much at home in tree crowns as on the
ground. Its coat is soft and luxurious.

# CHAPTER FOURTEEN

# PINE MARTEN AND FISHER

In some areas, the pine marten is less dependent on red squirrels for its diet than was earlier thought. It does depend upon them for shelter—sometimes expropriating the tree hollow of the same squirrel that served dinner.

Of all the mustelids the pine marten, or American sable, is the one I know best, while the somewhat similar fisher is the species I know least, having briefly seen only one in my lifetime. Both are of the same genus, *Mares*, the main difference being that the fisher averages two to three times larger than the marten. Their northern ranges across the continent greatly overlap, but I believe the fisher was originally more of an eastern species, while the marten may always have reached greatest abundance in the western part of North America. Both animals are tree-dwellers, whereas all of the other mustelids must be classified either as land-dwelling or largely aquatic.

My familiarity with martens comes from frequently sharing their company along the hiking trails of Grand Teton National Park, and from their frequent visits to the bird or squirrel feeders of several of my neighbors. Generations of one marten family have lived and raised annual litters of young in peace and safety beneath the floor and among the roof rafters of my friend's log-cabin home. The vicinity of this cabin has been an ideal place to observe the animals any time of year.

*Martes americana* is long, lithe, and fairly quick in all its activities. The normal color is a rich brown shading to darker brown or black on its feet and tail. The pointed faces of the older ones especially may show a tinge of gray, with short dark lines extending upward from the inner corners of the eyes. A particularly striking feature of this species is the bright orange throat and breast patch, though this isn't always visible on the busy, moving animal. Some individuals have only a pale buff throat patch.

A marten's fur is luxuriously long and soft. It is among the most valuable of all wild animal skins, a fact that has not worked it its favor. Trappers assure me the animals are very easy to capture. The tail is well-furred, and is by far the best register of its emotions and indicator of what the animal will do next.

Nose to tip of tail, this American sable will measure about two feet long when fully grown, with males slightly larger than their mates. A large male will weigh from 2½ to an extreme of four pounds.

What has impressed me most about martens is their great strength and determination relative to their small size. Once, to photograph a young marten that always hid under my friend's house when I approached. I stuffed a road-killed pine squirrel deep into a cavity in a partially rotted deadfall. I thought this might entice the marten to come out and investigate. The marten did more than that. It ran over the toes of my boots to get at the deadfall, which, with loud growling and clawing, it soon split. The animal then ripped apart and ate the squirrel only a short distance away, keeping a bright beady eye focused on me—or perhaps on its own image in the lens of my camera.

To tell the truth, it has taken considerable equanimity on the part of my friends to abide a family of martens sharing their home. Often in the middle of warm summer nights they have been kept awake by the noisy vocalizing and antics that are part of the marten mating season. Then beginning nine months later, my friends are entertained by the kittenish sounds of hungry baby martens wanting to be fed, or later by the kits fighting with one another over food. The relatively long gestation period for so small a mammal is a result of delayed implantation of the fertilized egg in the female uterus. Litter size is from one to five.

The only quiet time of year around that cabin is autumn, when the young of the year are finally grown and driven away by the parents to shift for themselves. That means each one will have to find a territory of its own, including a home in a hollow tree that will probably be appropriated from a squirrel.

Most of the martens Peggy and I have met have been in relatively dry areas of the lodgepole pine and Douglas fir forests common to northwestern Wyoming. But we have also found a few during midsummer at higher elevations close to timberline, or at least where forests are becoming thin at 10,000 feet and beyond. Outside the uncut stands of forest in national parks, martens may be found occasionally in slash or second-growth timber, but rarely in the clear-cuttings that are devastating so much of the West.

I am still puzzled by one encounter with a marten in midsummer of 1980. Peggy and I were hiking a trail that paralleled a Grand Teton stream at about 7,000 feet elevation. We stopped to rest beside the steeply tumbling stream and there noticed a dipper in the water. Dippers are small, dark-gray birds that live along cold, unpolluted waterways where there is enough insect life to sustain them. Of all native passerines, they are the only species that can plunge

Logging has proven particularly devastating to martens. It eliminates trees
that martens need for winter hunting as well as attendant prey. Logging also
removes most of the hollowed trees that provide marten den sites.

completely under water to gather insect larvae and
tiny aquatic creatures. We soon noticed that after
emerging from beneath the water with a beakful of
food, the dipper flew directly to a nest it had built
on the face of a large, streamside boulder.

We moved a little closer to the nest and could hear
the squeaking of chicks inside, even above the rush
of the stream. Suddenly there was a marten on the
scene. As we watched, it walked on top of the boulder
past the nest barely six feet away. If it noticed the
nest or heard the noisy chicks inside, it gave no in-
dication at all. Why would a wild pine marten, always
hunting and constantly hungry, miss such an easy
meal? Martens normally pillage any active bird nests
they come across. Our only explanation is that this
marten may have been very young and inexperienced.

Much of the early writings about martens state that
they depend heavily, if not mostly, on red or pine
squirrels for daily fare. But naturalist Adolf Murie was
the first to note, in his *Mammals of Mt. McKinley
National Park*, that relatively few red squirrels were
ever captured. Instead, those Alaskan martens live
primarily on meadow voles and red-backed mice. They
also eat ripe blueberries in summer and frozen ones
in winter.

After his observations from Alaska to northwestern
Wyoming, Murie found that blueberries, haw, and
mountain ash berries were often the choice of martens,
even when voles and other foods were plentiful. A
marten wandering high in the Wyoming Rockies will
catch an occasional pika, golden-mantled ground
squirrel, young snowshoe hare, and ruffed or blue
grouse. Young grouse may be especially easy to catch
on the ground. Altogether Murie documented 120
food items of this species.

Over at least part of the pine marten's range, the
red squirrel, or chickaree, can be important prey.

There is evidence that martens migrate seasonally, up and down, from high-elevation summer habitat to lower altitudes when deep snows begin to accumulate. That's when a marten is most likely to move into an old cabin, line camp, or outbuilding as a semipermanent shelter, rather than travel along a circuit of summer squirrel nests.

During a recent November, my friend Art Smith went elk hunting late in the season to stock his freezer. Toward dusk one cold evening he shot a small bull and hurriedly field-dressed the carcass in the waning light. He planned to return early the next morning with a horse to pack out the fresh meat. But soon after daybreak he found that an angry, snarling pine marten had gotten to the carcass first, taken possession, and was reluctant to relinquish the prize.

If you think it strange that a three-pound creature would try to claim a meat supply from a 190-pound man, take my word that it isn't. This wasn't the first time for such an incident. In Colorado a particularly aggressive marten moved in to camp with outfitter Harry Curtis, his guides, and his hunting client. The marten slept in the woodpile, but didn't mind walking through the middle of the cook tent on its way to feed on quarters of elk hanging on the game pole outside. It would eat strawberry and raspberry preserves, but didn't care for the cherry variety. When the hunting season finally ended, the crew tried to lash the last haunch of venison onto an already nervous packhorse. But the marten rushed in a fury at the men, spooking the horse in the process. So Curtis had to strike a compromise with the small creature or face spending the winter in the high country. He left half the meat for the marten.

The more often I mentioned writing this book to neighbors, the more interesting marten stories I heard. One lady who lived near Alpine, Wyoming, was also an enthusiastic feeder of birds throughout the winter. But like so many others, she was plagued by larger birds such as ravens, magpies, and Clark's nutcrackers. They monopolized the place, driving the smaller and (to her) more desirable species away. So she designed an elaborate heavy-wire feeder that nothing larger than a grosbeak could use. Her feeder even discouraged red squirrels, which some say is almost impossible. Then one morning she heard what she thought were several cougars in a furious fight. Actually her problem was just one pine marten that had managed to enter the feeder and was in the process of tearing it apart to get out. After that, the woman simply fed her birds on a tray.

Pine martens may reach a fairly old age in good habitat where food is always abundant. One male, distinguished from the others by a light mark on its throat, lived around an Alaskan cabin along the Yukon River for 15 years, according to a homesteader and trapper there. There is also a record of another living 17 years in captivity.

In one of the few marten studies ever made, Dr. Tim Clark and Tom Campbell live-trapped the animals in two separate areas of northwestern Wyoming. In the Hawk's Rest area of the Teton Wilderness the average age of live-trapped animals was 3.6 years. And on the west slope of the Teton range, the live-trapped martens averaged only 1.2 years of age. The team determined age by counting the number of layers of dental cementum on the teeth.

Dr. Clark's investigations have convinced him that martens quickly leave areas of logging and especially clear-cutting—as on the west slope—where all the timber of an area, marketable or not, is removed. The practice eliminates downed or leaning trees, which martens need for winter hunting. Most modern timbering also removes dead and "over-mature" trees that have hollows and are perfect for denning and resting sites. When those are gone, so are the martens—as well as all the many species of birds and squirrels that rely on such trees and the insects that soon invade them.

**THE FISHER.** The marten may be fleet and agile enough to occasionally catch a red squirrel in the treetops. But even faster in the crown of a forest is the fisher, *Martes pennanti*, which has been known to catch and dine on the marten, its close relative. Although the fisher may be the fastest of all native omnivores in the trees, it spends most of its time on the ground, where it can outrun a snowshoe hare. Its legs are short, but it can cover 16 feet or more in one bound.

For an exciting, hopeful time in the early 1980s there were many reports of seeing eastern cougars once again in the wilder parts of New England and New York State. The animals were spotted briefly along hiking trails and crossing the roads at night. But when conservation personnel were called to the scene for verification, they usually found the tracks not of a cougar, but of a fisher.

The mistaken identity is easy to understand. An average fisher is two to three times as large as its cousin the pine marten. It is dark brown to almost black, and very old individuals may be grayish around the neck. The fisher has a short broad head, short ears, and a long tail that arches out behind as it bounds along. At a distance this movement is catlike, and from afar and in dim light the animal may appear much larger than it is. In reality fishers weigh from 5 to 15 pounds, and the male is more than twice the

size of the female. Among the largest on record is a male trapped in 1972 near Sebec, Maine, that weighed slightly over 20 pounds.

A full-grown fisher has no natural enemies. Starvation is rare among the animals, and disease even rarer. But trappers complain almost as much about fishers as about wolverines. Like those other mustelids, fishers tend to follow traplines at night and strip them of the catch. For the most part, however, the fisher is a recluse that few people know and most people are totally unaware of despite its extremely vauable fur.

How the fisher got its name is another unsolved mystery. Also known as *pekan*, *Pennant's Cat*, and *fisher marten*, to list a few, the species may have been confused by early American settlers with the otter, which is the fisherman of the mustelid family. They may also have confused it with the European polecat, which was once called a *fitcher*. But no matter. The species does not prowl streams and lakeshore to find fish. Rather, it is a resident of wooded ridges and coniferous forests.

Fishers were probably quite numerous in what is now the northeastern United States and eastern Canada when Europeans began to settle the area. The animals were still common in the annual harvest of trappers in Maine, Nova Scotia, and New Brunswick in the late 19th century, although pine martens were caught more often. It was during that time that the incredible going price of $200 (the equivalent of $4,000 today) for a single, silky female pelt almost eliminated the species from its territory. This was an era of rural sustenance living, when most citizens depended on seasonal trapping for their only cash income. Trappers were unrestricted by bag limits or closed seasons, and game management was a concept yet to emerge.

One fatal habit often made the fisher its own worst enemy. The large mustelid would relentlessly follow the same 20- to 30-mile hunting circuit, week after week. Once an observant trapper found a fresh track, he knew just what to do. By about 1900 though, trapping was no longer efficient in time or cost. So trappers turned to tracking females over the snow, often with dogs bred for the work. With sufficient provisions in a packbasket for a week or so in the woods, a fisher hunter would simply walk until an animal was treed or chased into a hollow log, from which it was shot or smoked out. Hounds that could trail and tree fishers like coonhounds were worth more than gold. At then-current fisher pelt prices, a top quality hound could

Faster and more agile in the forest than the marten, the larger fisher easily captures snowshoe hares on the ground. An adult can cover about 16 feet in a single bound.

This indolent pose belies the speed and ferocity with which a fisher can attack its prey. In fact, the fisher is one of the few predators that can kill a porcupine without being badly wounded with spines itself. This feat is accomplished by repeated, lightning-fast attacks to the porcupine's face.

buy its owner a new house in only one season, according to the notes of a contemporary northern Maine fur trader.

Hunting and trapping were only partly to blame for the vanishing fishers of the East. Habitat destruction was the main factor in their decline. By the 20th century, only a fourth or less of the original New England and Maritime Canada forest was still uncut. Most had been logged and cleared for farming and grazing livestock. Forest fires fed by evergreen slash raged out of control. During the two summers of 1903 and 1908 forest fires burned a million acres of fisher and other wildlife habitat in the Adirondacks alone. It wasn't until 1929 that Vermont became the first state to protect the fisher from hunting and trapping. Maine followed six years later. But fishers were already extirpated from all but a few of the most remote areas.

During the Great Depression and the period between the two World Wars, the human population shifted from rural areas to eastern cities. Some farmlands slowly began to revert to woodland. But it was a prey species, the porcupine, that sparked man's first effort to restore fishers to their old haunts.

By the mid-1950s porcupines had become a plague in New England. They ate axe handles and canoe paddles, gnawed on outhouse seats and vacation cabin doors; but worst of all, they girdled and killed the softwood trees that had become important in a growing pulp industry. In 1957 the Vermont legislature appropriated $5,000 (then the price of a few fisher pelts) for a porcupine-control project. Over a hundred fishers live-trapped in Maine were released in 40 Vermont areas where extensive damage was evident.

Apparently the project worked. Soon porcupines seemed less visible, and by 1974 there were enough fishers stalking the Vermont woods to justify the first trapping season in almost half a century. Fisher populations did increase in other New England states at the same time, however, and without any restocking attempts. The price of a prime pelt in 1985, incidentally, was back down to $200, the same amount of a century ago.

One reason fishers will probably never be abundant, even in the best environment, is their low reproductive rate. Breeding takes place in late winter or early spring. Two or three kits are born almost twelve

This dark predator is an especially effective hunter of raccoons. Raccoon numbers are small in fisher country. However, squirrels and other small mammals are the preferred prey, and studies of stomach contents show that fishers also feed on birds, insects, mosses, ferns, fruit, fish, and even beechnuts.

months later in a hollow tree or log. The long gestation period, which exceeds that of black bears, results from fertilized eggs remaining dormant in the mother until near the end of those 11-odd months. The female comes into estrus again only a week or two after giving birth. If a male exists anywhere in the vicinity, the female is immediately bred again.

There is little data about how young fishers develop and how they learn to hunt and cope with life on their own. But families usually disintegrate by early winter, and immature fishers are forced to find their own hunting territories.

A common myth involves the method by which a fisher kills and eats porcupines. The long-held belief was that somehow, dodging the spiky armor, the fisher got a paw underneath and flipped the porcupine over on its back, laying bare its unprotected throat and belly. But we now know better. Maine researcher Dr. Malcolm Coulter discovered that the fisher simply bores in from the front with astounding speed, repeatedly striking the prey in the face until it is weakened enough to be rolled over. The porcupine's hide is so neatly peeled from the carcass that the job appears to have been done with a sharp skinning knife.

The fisher may acquire a few quills in the scuffle, but these usually work free in time. According to naturalist Victor Cahalane, "Apparently the porcupine quills . . . . often go through the outer hide without piercing the second layer of skin, or the muscles. Instead, they turn until they are flat against the skin and work their way out again. This usually occurs without causing any inflammation or festering sores. Quills have been known to pass through its digestive tract without piercing the delicate lining. Fishers are not always invulnerable, however. A number of them have been found with faces studded with quills and dying of starvation."

Though fishers may kill a winter-weakened whitetail deer or a spring fawn as well as a porcupine, smaller game is its usual fare. As noted earlier, the fisher rarely eats fish except those used to bait traps for mink. Researchers Art Cook and W. G. Hamilton, Jr., analyzed the stomach contents of 60 fishers in the Adirondacks of New York and found the remains of ten different mammals, as well as of birds and insects, mosses and ferns, fruit and fish, and beechnuts.

The stories about the potential threat fishers pose to people may largely stem
from reports of trappers who have encountered angry fishers in their traps.
Fishers normally try to avoid people.

Squirrels are preferred prey, however.

Fishers are also among the few effective predators of raccoons; thus it is no coincidence that numbers of the latter border on nonexistent in fisher country. The fisher can as easily catch 'coons in trees as on the ground. There have also been reports recently of house cats disappearing in northeastern areas where five-toed fisher tracks are common over the snow.

Essentially this dark predator is a solitary animal, with a reputation in some places of being a very aggressive, fierce animal to be avoided. That may be a result of stories about animals caught in traps. For example Leonard Lee Rue III, noted naturalist and wildlife photographer, tells of one trapper who was bitten by a fisher "all the way through his boots, pants, and long johns." Another fisher shredded the ear of a hound dog that tried to dig it out of a rotten log. Most fishers simply want to avoid people, and I have heard stories that they have even refused to cross the fresh tracks of a man on snowshoes.

If you are ever able to sight a single fisher at close range in your lifetime, you will be lucky indeed.

Few, if any, animals wear warmer, more luxurious fur coats than the Ameri- ▷
can mink. Sleek and slicked
back in water, the fur is remarkably lustrous when dry.

# CHAPTER FIFTEEN
# MINK AND WEASELS

Mink country is normally moist and well watered, and is also home to beavers and muskrats. Mink travel on trails that parallel waterways and also lead from one lake to another.

For centuries the mink of Europe and later those of North America, *Mustela vison*, have furnished women with one of their most visible status symbols. To my view few coats of any kind are as beautiful and glamorous, or as practical and utilitarian, as mink coats—provided, however, that they are worn by their original owners.

When the mink swims, its fur slicks back and becomes a surface as smooth as a rainbow trout's. On first emerging from the water, a mink still appears as sleek and slippery as an eel. But moments later out on the bank, one quick, tremorlike shake of the body and every hair stands up free and seemingly dry to reveal the softness and color of the magnificent pelage. To see this transformation just once, and to watch the antics of a wild mink, is an extraordinary outdoor experience. Unfortunately we don't encounter the animal often enough.

The mink is the largest North American member of the genus *Mustela*, which includes the weasels and the black-footed ferret. Northern males reach a length of two feet or slightly more, about a third of which is accounted for by the bushy tail. Northern females are a few inches shorter, while mink of Louisiana and elsewhere in the South mature slightly smaller than those of the American North. All are longer and heavier in the body than their closest relatives, the weasels. In fact *water weasel*—along with *minx* and *belette* in Louisiana—is a common name for the species.

Wild mink are a uniformly dark brown above and below except for occasional faint white blotches on the chin, throat, chest, and abdomen. The brown of the upperparts varies from burnt umber to chocolate, while that of the legs and underparts is a little lighter in color. The tail is the same dark brown as the back at its base, but darkens to almost black at the tip. A mink's tail is bushier and its entire pelage more dense and glossy than that of any weasel. Thick underfur and long, somewhat oily guard hairs are what make the animal's coat water-resistant and allow it to live a semi-aquatic life.

A mink's summer coat is paler and thinner than the prime coat it wears in early autumn. Except for the pads and semi-webbed toes, the feet are also densely furred. The anal glands typical of all mustelids secrete a substance with a strong, fetid odor, particularly during the breeding season and when the mink is aroused or excited. Many who trap mink or are familiar with them regard the odor as more noxious than a skunk's.

True color phases are unknown among wild mink. Very infrequently a blond mutation or an albino is seen, but the autumn haze, ambergold, sapphire, gunmetal, and other colors of mink coats found in furrier shops are products of selective breeding on fur farms across America. Of all wild mink, 99.9 percent are a rich dark brown, whether they're found in Florida or northernmost Yukon.

Except perhaps for the river otter, which is not nearly as abundant, the mink has the widest range today of any of the mammals described in this book. On the entire North American continent, the species is missing only from a corner of southwestern United States, from Mexico, and from Arctic islands north of the Canadian mainland. The mink is among those species that has adapted fairly well to civilization.

Mink are never found far from water and are most often seen in it or along its edge. Comparatively high populations live in usually humid southeastern coastal marshes. The tupelo gum swamps along the lower Mississippi River and the Atchafalaya Basin of Louisiana are excellent habitats. In Louisiana's St. James Parish alone, between 10 and 14 thousand mink are trapped each winter. The abundance of the species in the bottomland swamps there is due directly to the prevalence of crayfish, one of the mink's main foods.

From 1914 until 1973, no fewer than 4½ million mink were recorded taken by Louisiana trappers. No other state has approached that great a harvest.

One of my closest observations of mink took place far from Louisiana. It was in late January on frozen Sandusky Bay along Ohio's Lake Erie shore. Along with many others, I had been fishing through holes bored in the ice several hundred yards from shore. At dusk I gathered my gear to leave the ice and head homeward. That's when I saw a dark, low-slung animal stalking among the now-abandoned fishing sites, looking for bait or fish, or the entrails of fish left behind by fishermen. I paused and stood motionless while the mink came almost to my feet and peered into my bait bucket, which unfortunately was empty.

We found this mink litter in a hidden, grassy nest in a northern evergreen forest. The animals are born blind and nearly helpless.

Much has been written and claimed about the ardent though violent love-making of *Mustela vison*. But the animals are no more active in this respect than, say, cottontail rabbits, cougars, or whitetail deer. The mink mating season in Michigan occurs from late January into early February. The male is promiscuous and, like the rabbit and deer, spends that time finding and breeding as many females as he can. Competition with other males is probably intense. Females are deserted immediately after breeding to raise the young on their own, usually in a birthing den that is near, if not actually on, the water.

The young are born five or six to a litter in April or May in Michigan. At birth, each is the size and shape of a human's little finger. They are blind, naked, pale in color, and are totally dependent on the mother. After the family finally breaks up in late summer, the female resumes the same kind of solitary life her forgotten mate has enjoyed all along. She tends to live in a more restricted area, however, not too far from a more or less permanent den. This territoriality has made females, whose pelts are superior to male skins, more susceptible to trapping than their mates.

Many of the mustelids tend to be specialists, but the mink has evolved with composite talents. Though a mink can neither swim as well as an otter, climb as well as a marten or fisher, dig as furiously as a badger, nor catch mice with the deadly efficiency of a weasel, it does a commendable job at all of those. That may help explain why minks, like coyotes, red foxes, and raccoons, have been able to cope with civilization so much better than the specialists. Depending on which natural enemy may pursue it, a mink can escape by climbing, swimming, or racing for a den. If none of those is practical, or if one method fails and the animal is cornered, it can depend on sheer rage and savage counterattack. For its size, this solitary anti-social carnivore is an extremely strong and tenacious fighter.

Although captive and ranch mink have been studied and restudied because of their immense commerical value, biologists have not been successful in learning many of the secrets of wild mink. One researcher told me that the species was too high-strung for making any comprehensive studies. He pointed out that a female mink will often kill her own young at the slightest provocation or intrusion near a den area.

Food habit studies in several localities reveal that

mink are not too selective and that their hunting is guided mainly by whatever is available. Around water, mink can live on fish and frogs, crayfish and turtle eggs, as well as muskrats. There is much disagreement on how well they actually catch live fish. Though trout-hatchery operators despise the animals, live-trapped wild mink placed in tanks where they could be easily observed were unable to catch trout. They are known, however, to feed very well on suckers that spawn every springtime in the shallow riffles of many American rivers. They may also be able to catch bluegills and some other sunfish, which are slower swimmers than trout and also spawn in shallow water.

When touring the countryside, the mink will kill small birds and eat the eggs. Other prey are mice, rabbits, probably snakes, and certainly domestic poultry wherever available. Farmers with chickens hate mink as much as they do red foxes. But in many of America's vast marshlands, the muskrat is the mink's primary and occasionally sole prey. In the marshes along the southwestern shore of Lake Erie—where the density of both mink and muskrats has always been high—mink spend most of their active hours, day or night, in a tireless search for fresh muskrat meat. After dining on or driving away the occupants of a particularly large or suitable muskrat

I've found mink frequenting areas of beaver ponds and cuttings, capturing beaver kits and using beaver houses for their own dens.

house, a female will then use this cabin in the cattails as her own den.

Most predation is on young muskrats because they are the easiest to kill and, for several months in summer, are most abundant. But beginning with the onset of winter and eventual freezeup of the wetlands, a mink must hunt adult muskrats, which are determined and formidable rodents of similar size.

The natural gifts for climbing, swimming, and running that make a mink an effective predator also help it escape when preyed upon.

Early in the 1950s I hunted ducks with Rudy Golic in the marshes of Lake St. Clair, Michigan. Golic was a trapper as well as a waterfowl guide who practically lived year-round on the water. Twice, he told me, he had seen encounters between mink and muskrats, and both times the rodents won. The first time, Golic saw a mink in hot pursuit of a young muskrat that squealed in fear as it swam toward the safety of its cattail house. The mink caught the kit right at the entrance, but that proved to be a mistake. Two adult muskrats, no doubt the parents, appeared on the scene and attacked the mink from opposite sides. What followed was obscured by the roil of muddy water, but the mink was hard-pressed just to escape with its life.

On another afternoon, Golic sat in a reed duck blind and noticed a mink swimming an irregular course across the shallow bay. As it swam beside a driftwood log, a single muskrat emerged from a hollow and jumped down on the mink's back. That also was a furious, snarling, mostly obscured fight that ended when the mink managed to disengage itself and swim away. Golic recalled that the odor of mink musk remained heavy and unpleasant over the whole area for some time after the incident.

**WEASELS.** The mink's North American relatives are three weasels: the longtail, *Mustela frenata;* the shorttail weasel or ermine, *Mustela erminea;* and the least weasel, *Mustela rixosa.* The first two are similar in appearance, and their ranges overlap roughly along the U.S.-Canada border, in northern New England, the northern Rockies, and the Cascades. For the most part the shorttail is a northern species while the longtail lives in the lower 48 states.

My earliest weasel encounter occurred with one I never saw. It happened in southern Ohio when I was a young boy. Going out to my Aunt Frieda's chicken yard one morning to collect eggs, I found instead the bodies of two white chickens with their heads and necks pulled as far as possible through fine chicken-wire fencing. There were bloody punctures at the base of each neck. My uncle Ernst, who was normally unpleasant anyway, was now livid.

"A *bleep-bleep* weasel did it," he thundered.

Later I realized that a weasel was indeed the only animal that could have squeezed through the wire. The predator had tried to drag the chickens through the fence head-first, but of course it couldn't pull it off. All this was much easier to understand later on when I saw my first weasel, a longtail.

The shorttail weasel exchanges a brown and yellowish coat in summer (above) for pure white fur in winter (next page). As shown, the change to winter pelage may occur before snowfall.

That longtail weasel that I first saw was brown on the back and sides, light underneath, and sleek. But most remarkable was the slender, almost snake-like body and very short legs, especially the forelegs. Later I would learn that the longer hind legs are only about half as long as the body from the base of the head to the base of the tail. In raccoons, by contrast, the hind legs and body length are about the same. The short legs on such a long body account for the weasel's distinctive running gait. At every bound the elongated body arches upward, resembling an inchworm's movement. I marveled at this creature—a small thing, yet the only one that could so anger my Uncle Ernst and get away with it.

Like snowshoe hares, weasels that inhabit snowy areas turn white in winter. Even northern weasels that live entirely indoors or are captives in zoos change color because the hormones responsible for color are triggered through the eyes by the length of the day, not by falling temperatures or snow depth on the ground. I have seen all-white weasels running over the brown forest floor of a Wyoming November. And I have also seen brown ones well after the first snows fell in the same area. The color change is gradual, usually beginning on the belly and around the face and working back toward the tail. Only the black tip of the tail remains black in shorttail and longtail weasels.

Male longtails, the largest native weasels, may reach 18 inches in length and weigh from 7 to 12 ounces. A large male shorttail weasel may measure 13 to 14 inches and weigh 4 to 6 ounces. Females of both species are only about two-thirds the size of males in the same range and of the same age. Because they are so slender, all weasels weigh only a fraction of what other animals of the same length weigh.

One or more species of weasel can be found anywhere from northern Alaska down to northern South America. Weasels can even be found as high as 12,000 feet in the Colorado Rockies, and except in the driest portions of the American Southwest. Any longtail weasel living elsewhere in the Southwest will have a distinctive white pattern on its face and forehead, so that it appears masked.

Like all mammal carnivores, weasles evolved from miacids, an extinct group of mammals that lived on earth 35 to 50 million years ago. But they have retained many of the characteristics of miacids, including their short legs, solitary life-style and hunting skills. All North American weasels are well equipped for survival in this modern world.

Folklore about weasels originated in Europe and was embellished in the New World and has never been complimentary or accurate. Any weasel was regarded as a bloodthirsty, wanton killer. Even today an un-savory person who reneges on debts or his word is known as one who "weasels out." Many of the old ideas about the species even found their way into early scientific literature. But the latest studies have clearly shown that weasels, like many of the other carnivores, deserve a better image than they have had.

To better understand these animals, consider first the demands placed on them by their unique body shape. That is, weasels, like all mammals, are endothermic, which means that they maintain a constant body temperature, usually higher than that of the environment, by using metabolically produced heat. But because weasels are long and slender, they have a greater surface area than other mammals of similar weight but with more compact shapes. Weasels lose heat more quickly and thus need higher metabolic rates to compensate for the loss. Therefore, weasels must eat more often than other animals their size.

Because they burn off calories so quickly, weasels must take advantage of whatever protein source they can find. They will pass up nothing, from insects to the remains of animals killed by other predators, including deer and other animals killed by cars along the highway. I have found a surprising number of weasels that were also killed along paved roads when attracted there by the carrion. Live mice and chipmunks constitute most of a weasel's diet, however.

The weasel's small size in turn limits the size of the prey it can subdue. Hunger and determination notwithstanding, some smaller animals are too strong for a weasel. I've seen shorttail weasels capture half-grown cottontails. Yet the Uinta ground squirrels that have colonized our backyard—and scatter when a goshawk is overhead or a pine marten appears in the vicinity—do not seem unduly alarmed when the local shorttail weasel passes through. Recent studies under controlled conditions have shown that a full-grown cottontail rabbit may fatally injure a longtail weasel as often as the weasel can fatally injure the rabbit. Thus rabbits and hares are on the upper limit of a weasel's prey size. Weasels may or may not try to attack them, depending on how hungry they are.

A weasel's high metabolic rate means that each food item killed represents an investment in energy and has a high replacement cost. It is therefore not surprising that weasels defend their catches ferociously. In wilderness camps and cabins, I have had wild weasels try to take the mice I had removed from traps right out of my hand. I have known many other instances of the small mustelids boldly confronting humans for a piece of meat.

Hunting behavior, too, is largely determined by the high metabolic cost of catching prey. In those rare times when a weasel finds, say, a litter of chipmunks

Short legs and a long body account for the weasel's distinctive running gate—with the body arching upward at every bound, resembling the movement of an inchworm.

in a snug den, a weasel kills as many as possible, gorges on the prey, then caches the rest for later. That accounts for the stories I often heard in my youth about a weasel killing all the chicks in a farmer's chicken coop.

Weasels caused plenty of headaches for one biologist who was trying to live-trap fishers on Michigan's Upper Peninsula. For this purpose he used large, wire-mesh box traps baited with strips of venison. But a weasel always found the traps before the fishers and ate the bait. That alone was no problem because the meat chunks were large enough for both predators. But the trouble was that the weasels would trip the trap doors on entering, feed, and then escape through the mesh. The biologist solved the problem by setting an extra set of baited traps for the weasels and then moving them a great distance away.

From time to time Peggy and I have had a shorttail weasel frequently pass through our property and check out the bird feeder area, which the grosbeaks and Steller's jays share with chipmunks. We enjoy seeing weasels and have never had a mouse problem as have some nearby rural dwellers. All three species of weasels

are extremely valuable mousers around country cabins and camps.

Weasels normally hunt by a strategy that might be called random search. They move quickly and constantly, changing directions often and searching out all the crevices, hollows, and other places where small rodents might lurk. They inspect everything. I've even seen them tunnel under winter snow and climb trees in summer. One morning Peggy and I saw a weasel running away from a nest with a rufous hummingbird in its mouth. The weasel's elongated body also allows it to squeeze down into mouse holes too small for anything else except snakes.

Weasels kill by rushing directly at the prey. If the prey is small, such as a mouse or chipmunk, the weasel tries to grab it by the head or back of the neck. That single bite is usually enough to kill it. But if the predator only manages a grip on a tail, leg, or flank, it will try to wrap itself around the mouse or chipmunk, gripping with all four paws. The weasel then releases its initial grip and turns the prey so that it can reach and bite the back of its neck. All of this takes just seconds. More often than not it takes place at night.

The least weasel is the smallest living carnivore, with the males measuring
just six to seven inches in length and weighing only about two ounces.
(Kark Maslowski photo)

The hunting habits of weasels have recently been the subject of some interesting field studies. Weasels are estimated to sometimes travel four times the distance of a straight line between two given points, and there are certain advantages to that. Not only can a weasel thus inspect more places thoroughly, it may also confuse potential prey. A mouse, for example, is never certain which way to run for escape. The irregular travel also makes it difficult for a larger predator to zero in on the busily hunting weasel. In addition, the prominent black tail tip may cause a raptor to strike too far back for a secure grip.

Small as they are, weasels are fair game for many species including closely related martens, fishers, and wolverines. Bobcats and lynx are natural enemies. Weasel skulls have often been found among the pellets beneath great horned owl nests. Goshawks catch them and, for that matter, so do gyrfalcons, prairie falcons, and roughlegged and redtailed hawks. A large, feral house cat killed a weasel that had long lived around the cabin of a close friend of mine in northern Wisconsin. I am happy to report that my friend promptly dispatched the cat.

Although weasels are no more and no less important than any of the other creatures that share their natural ecosystem, they are tremendously valuable to ranching and farming in many parts of America. Across the western United States for example, pocket gophers are a persistent nuisance because they crop the alfalfa planted in irrigated fields and compete with livestock for the native vegetation in mountain meadows. By coincidence there, pocket gophers are the favorite prey of longtail weasels.

**THE LEAST WEASEL.** So far this chapter has concentrated on the more familiar shorttail and longtail weasels. The least weasel, though common in some areas, is more furtive and seldom seen. Resembling an elongated mouse, a female least weasel is the smallest living carnivore. It measures five or six inches overall and weighs less than two ounces. Male least weasels average about seven or eight inches long, including the tail, and weigh two ounces, or less than some of its prey. Least weasels do not have black tips at the end of their tails as the others do.

Least weasels feed almost entirely on mice, and often take over a mouse nest as their own. Three to ten young, though usually four or five, are born in a litter at any time of year; and there may be more than one litter every 12 months. The color of a least weasel's fur also changes to white in winter, except in the American South.

Black-footed ferrets may now be extinct from the wild. Habitat destruction, loss of its prey base, and ▷
disease can all be blamed. (LuRay Parker, Wyoming Game and Fish Department, photo).

# THE BLACK-FOOTED FERRET

Late one night in September 1981 a dog named "Shep" belonging to rancher John Hogg of Meteetse, Wyoming, killed a weasellike animal that may have been trying to eat from a dish of dog food. Hogg was persuaded by his wife to take the animal to Larry LaFranchi, a taxidermist who lived close by. LaFranchi concluded the unknown creature was a black-footed ferret and immediately notified the Wyoming Game and Fish Department in Cheyenne.

The news quickly spread all over North America among those concerned about wildlife and the environment. That isn't any wonder. Despite a handful of scattered sightings in recent years, none of them confirmed, most scientists grimly dismissed the ferret as another native American species that had vanished forever from the face of the earth. Professor Tim W. Clark, an Idaho biologist who maintained a long-standing interest in the species, was among the first to rush to Wyoming for an on-the-spot appraisal. Maybe the ferret, *Mustela nigripes*, wasn't extinct after all.

Very little had ever been known about the black-footed ferret. The first illustration of it was painted in about 1844 by James W. Audubon, son of the premier bird artist. In fact the first ferret pelt acquired by younger Audubon and naturalist John Bachman near Fort Laramie, Wyoming, was the only known specimen for many years. When the pelt was inadvertently lost, some naturalists suspected that Audubon had simply invented the animal.

**PHYSICAL CHARACTERISTICS.** We do know that black-footed ferrets, unlike the closely related mink and weasels, are specialists. They live in prairie dog burrows and prey almost exclusively on prairie dogs. Black-footed ferrets have long bodies and short

These night photos of the black-footed ferret were among the last taken of this animal in the wild. As this is written, scientists are laboring intensely to help a small captive group of ferrets increase their numbers sufficiently to allow the return of ferrets to the wild. (LuRay Parker, Wyoming Game and Fish Department, photo).

The ferret has preyed almost exclusively on large prairie-dog populations on wide-open western plains. With this near total dependence on a single prey species, disease vectors affecting the prey have proven devastating to the ferrets.

legs suited to exploring underground tunnels. True to their name they have black feet, as well as black-tipped tails. Their bodies and short tails range from yellowish-brown to buff-colored. Ferret faces are distinguished by black, raccoonlike masks. Because ferrets live below ground and are nocturnal, they are relatively safe from such natural enemies as raptors, badgers, and coyotes. But it also has kept their presence and habits secret from man.

Tim Clark immediately searched the known prairie-dog colonies in the Meteetse area where the ferret had been killed. He discovered prairie-dog towns unknown until then, but no black-footed ferrets. No sooner had he returned home to Pocatello, Idaho, when a cowboy on a ranch near John Hogg's reported that he had seen a live ferret while working cattle. Furthermore, the cowboy said his dog had chased the ferret into a hole from which it bobbed several times while he had studied it from only a few feet away. The ranch hand then guided biologists to the exact area.

That's how on a raw morning in October, 1981, Steve Martin and my old friend Dennie Hammer of the U.S. Fish and Wildlife Service became the first humans in many years to hold a live black-footed ferret in their hands. Few of the mustelids had ever been captured alive. Martin and Hammer radio-collared the animal, which soon led the investigators to other ferrets. Eventually they observed nine in the area. It was beginning to appear as if a lost species was returning from the dead.

On a bitter December night two months later Tim

Clark had his own first look at a live ferret, and by that Christmas a total of at least twelve different ferrets had been sighted.

**FROM THE BRINK OF EXTINCTION.** As late as 1877 when the second ferret pelt was discovered near Cheyenne, prairie dogs still lived in inestimable numbers all across the western Great Plains from the Texas Panhandle north to southern Alberta, and from Utah eastward to Nebraska. Naturalist Ernest Thompson Seton figured the total population to be about five billion. Certain prairie-dog towns were believed to stretch for a hundred miles in all directions. Theoretically such a multitude of prey could have supported tens of thousands of ferrets, but considering that pioneers traveling and settling the West mentioned them so seldom, ferrets may never have been abundant.

Disease or epidemic may also have led to the ferret's radical decline, though the deliberate attempt to eradicate prairie dogs was probably the main factor. Everywhere the prolific rodents were considered terrible range pests, as bad as a plague of locusts. Homesteaders tried to get rid of them by plowing under their towns. When that didn't work, they turned to strychnine, cyanide, dynamite, and buckshot. All across the Great Plains we did eventually almost wipe out prairie dogs just as we had wiped out the great herds of bison. It is believed that the area of prairie-dog towns surviving in the West today is less than one percent of its original extent. Of course the black-footed ferrets that preyed on them were all but eliminated in the process.

Northwestern Wyoming was one of the few areas where ranchers did not energetically try to eradicate prairie dogs. For generations, most ranchers there had a live-and-let-live attitude, and that contributed to the ferret's survival. Tim Clark found that many landowners were interested in the ferrets' continued existence, and some were very cooperative during the scientific studies that followed.

**BEHAVIORS.** During the typically severe winter of 1981–82, when the temperature once dropped to −38°F, Clark and several colleagues camped out in black-ferret country. Many mornings their boots were frozen to the ground and they cooked breakfast on an iron kerosene stove with stiff, icy fingers. They hoped for light powder snowfalls during the night so that in daylight they would not mistake old tracks for fresh ones. In time they were able to count the tracks of 22 different ferrets in one five-square-mile study area. The location of the region was kept secret. Clark correctly felt that any unnecessary human activity in the vicinity might well affect the survival of a gravely endangered species.

The first conclusion of the winter's observations was that ferrets are solitary animals. The telltale tracks revealed that individuals hunted alone in their own territories. This pattern changed in early spring, however. Tracks began to overlap and nightly travel distances greatly increased. Clark realized that the animals were now out hunting mates as well as (or instead of?) food.

A grant from the National Geographic Society enabled Tim Clark and his team to eventually chart all of the 21 prairie-dog colonies believed to contain ferrets. All of this work was done on foot. The men walked about 2,500 miles and counted 111,000 dog burrows in 7,000 acres of towns. The grueling effort seemed worthwhile when during the night of June 28, 1982, an adult ferret was seen carrying three tiny kits, one at a time, from one burrow to another. Another night associate researcher Steve Forrest counted the green eyes of 16 different animals in his light beam. That is probably the most ferrets any human being had ever seen in such a short time.

By the time summer was blending into Wyoming's early fall, Clark's group had identified 12 litters. They also estimated that at least 38 new ferrets had been added to the local population, an astounding success for the species.

From all the above, conservationists had a right to be guardedly optimistic. Maybe this unique and rare carnivore really *had* come back. The Wyoming Game and Fish Department was enthusiastic and became more and more involved. Geneticists noted that a population of about 500 individuals would be needed in the Meteetse region of Wyoming to bring the species to a safe, healthy status. And that didn't seem impossible at all. But what followed is not a happy story.

As interest across America increased and more biologists became involved in ferret investigations, more and more of the animals were suddenly found. By the end of summer 1984 a live-capture and marking project indicated a population of at least 128 ferrets living in the Meteetse area. That was the crucial point at which state and federal officials could easily have taken bold action toward permanent ferret survival. But they missed the boat.

Biologists who had worked on ferret studies told me that the high number of ferrets was actually due to temporary overpopulation brought on by a series of unusually successful breeding seasons. It was thought that this number could not be maintained in a limited territory. A lot of the animals were certainly going to die. Thus, that would have been the perfect

The lessons learned from the efforts to save the last few black-footed ferrets should prove invaluable in efforts to save other species headed toward extinction. (LuRay Parker, Wyoming Game and Fish Department, photo)

time to live-trap some of the surplus and move those born-in-the-wild animals to restock areas where they formerly existed.

As it happened, the black-footed ferret's original range included such sanctuaries as Devils Tower National Monument in eastern Wyoming, as well as Wind Cave National Park and Custer State Park in South Dakota—all less than a day's drive from the Meteetse region. All had good numbers of prairie dogs, and all had the same high-plains weather and vegetation so familiar to the ferrets. But no such restocking action was taken, and twelve months later the population had fallen from 128 to 60. By October 1984 there were only 31 known ferrets left alive.

**DISEASES.** Canine distemper was diagnosed as the cause of the decline. It may have raged through the population in any event, but it was almost certainly spread more rapidly by the unnatural overpopulation. By late 1985 the number of surviving wild ferrets was estimated at from only a dozen or so down to five or six. As a result, a decision was made to capture the last ones, or at least as many as possible from the dwindling Meteetse population, and propagate them in captivity.

The decision was questionable at best. In 1970 a small colony had been found in South Dakota. Giddy from minor successes in propagating some other endangered species, the U.S. Fish and Wildlife Service had decided that black-footed ferrets simply couldn't make it on earth without captive breeding—even though there had never been a successful laboratory breeding of the species. They live-trapped six ferrets, of which four soon died of canine distemper. The four died when they were infected by innoculating them with distemper vaccine. Three more ferrets were then captured in South Dakota and all survivors were taken to the Patuxent Wildlife Research Center in Laurel, Maryland, half a continent away from their natural climate and surroundings. There one of the females twice delivered litters of five young, none of which survived very long.

Even after that bitter South Dakota precedent, the Wyoming Game and Fish Department still live-captured six ferrets in the fall of 1985. All of these also soon died of canine distemper. From the beginning captivity, all the ferrets were kept in a single room. One animal that had brought the disease in from the wild passed it on by coughing and sneezing.

Following this fiasco, an immense amount of criticism was heaped on Wyoming officials, some deserved, some not. State biologists managed to capture six more, which were also taken to Wyoming's Sybille Wildlife Research Unit. Individuals were now separately quarantined and, as of this writing, they are still living in the seclusion of a sterile laboratory.

Few visitors ever even see the Sybille Six because it has been found that the ferrets are susceptible to influenza and other diseases transmitted by humans. Those who do see the ferrets must first scrub thoroughly in a shower room. They must also exchange their clothing including underwear for an antiseptic jumpsuit before entering the breeding area to see animals that are surprisingly small and attractive.

Another serious threat to ferret survival was discovered when in 1985 a Guatemalan bat researcher found that prairie dogs of the Meteetse area carried sylvatic plague. This is a bacterial disease in wildlife transmitted by fleas. It is called bubonic plague when humans contract it. Although we might assume that natural predators of prairie dogs would acquire a natual immunity from sylvatic plague, black-footed ferrets apparently do not.

That stimulated still another disagreement in Wyoming. BIOTA, the research group of biologist Tim Clark, called for a predator-reduction program covering badgers and coyotes, weasels and hawks, owls and eagles in the Meteetse area to reduce the possibilty that the ferrets would catch this newly found disease. The Center for Disease Control in Fort Collins, Colorado, in turn called for a blitz of the Meteetse ferret area with DDT, a particularly toxic, condemend insecticide that in my opinion has no justifiable use.

Fortunately the Wyoming Game and Fish Department turned down both the predator control and DDT proposals. Unfortunately, however, they did buy 6½

tons of a relatively benign insecticide, Sevin, to poison the prairie-dog fleas. It was hand-sprayed down about 100,000 prairie dog burrows in what may be the greatest plague control effort ever made in America. More likely it was about as worthwhile as pouring money down rat holes.

There remains this ironic possibility: The Sevin dusting, which was done more for political than biological reasons, may have been undertaken when the plague outbreak was already ending. It is even possible that by saving ferrets from plague, we may have given them canine distemper instead. Harry Harju of the Wyoming Game and Fish Department, who chairs the ferret advisory team, notes that the very people doing the spraying could possibly have spread distemper since they did not take the same precautions that regular researchers do before coming into contact with live ferrets. That is, they did not wear surgical masks, wash carefully, or avoid handling of dogs or other animals before entering the ferret area.

**THE FUTURE.** The future for this rare mustelid does not appear bright. Its best chance is to somehow survive in the wild around Meteetse, if a sufficient healthy population should still exist there. That's a big "if." Finding other populations and then moving any surplus animals to traditional areas is a remote possibility. And trying to raise young in a sterile laboratory and then restocking the wild with these sheltered animals seems to me to be a pipe dream at best. Additionally, current captives are believed to be closely related; thus if they do breed, genetic depression could doom the offspring.

On March 18, 1987, Bob Oakleaf of the Wyoming Game and Fish Department announced that a four-year-old male ferret had been live-captured and added to the 6 males and 11 females already in captivity. The biologist believed this new addition might prove extremely valuable to the captive breeding effort because it may be the only male of the lot with actual breeding experience.

Data on reintroducing captive animals, particularly predators, successfully into the wild simply does not exist. I can't help but wonder how the ferrets would fare today if that first one hadn't been killed by Shep and they hadn't been discovered by humans who couldn't leave them alone.

Unfortunately river otters fall into the too familiar category of disappearing wild creatures. Water quality, or lack of it, is probably the main reason.

# CHAPTER SEVENTEEN
# THE OTTERS

Adult river otters weigh from 15 to 35 pounds and
measure from three to five feet in length. Females
average about 25 percent smaller than males.

On a cold November day in 1980 Colorado biologist Dave Stevens stood frustrated atop an old beaver lodge that was almost completely obscured by drifted snow. All day long he had tracked a river otter through the snow, but a mile or so back he had suddenly lost its trail. Now the radio receiver in his pocket picked up a strong signal from a transmitter implanted in the body of that animal.

The otter seemed to be directly underfoot in the beaver lodge. But how did it get there without leaving a track? No wonder Stevens felt defeated. As a research biologist of the National Park Service, he had never studied an animal that had proved as confounding as the river otter.

Two otter species exist in North America: the river, or land, otter and the sea otter. Though the two are somewhat similar in appearance, they are not closely related. What they share more than anything else is a scarcity in numbers relative to their former abundance.

**THE RIVER OTTER.** The story and status of the river otter, *Lutra canadensis*, is a sad and familiar one. Once its range included almost every corner of North America. But as with too many other creatures, its range has decreased alarmingly during the 20th century. Colorado, where Dave Stevens found the otter in the beaver lodge, is one state where otters disappeared a relatively short time ago. For example, it wasn't until 1950 that they were considered missing from Rocky Mountain National Park, where they had once been numerous.

Several reasons have been offered for the otter's decline. Over-trapping for fur has been blamed with good cause. But probably more important have been the falling water supply, poison from chemicals con-

This is one of a family of four otters we saw catch eight spawning brook trout in a Wyoming beaver pond. This skill maddens some fishermen.

centrated in food fish, and the relentless encroachment of civilization on streams and wetlands. No one knows exactly how much each of those factors has contributed to the otter's decline, or in what proportion. But beginning in the 1960s and 1970s, conservationists everywhere realized that the time had come to try to reverse the trend. Now attempts are being made to restore river otters to their original range. Dave Stevens was part of the effort to reestablish the animals in parts of Colorado that would be ideal for them.

The Colorado plan was like those in some other states. Four areas believed to offer the best habitat were selected for reintroduction: Cheeseman Reservoir, the Gunnison River (above and below Black Canyon), Piedra River, and the Kawuneeche Valley in Rocky Mountain Park. The latter includes Grand Lake and its tributaries, an abundant water supply, plenty of fish, and beaver lodges. Forty-one otters live-trapped in Washington, Oregon, and Wisconsin were released in Colorado waters.

The results to date have not been encouraging. Of the ten otters released in the Piedra River, some may still survive because there have been scattered sightings and other sign. But elsewhere it seems that the introduced animals have vanished. Wild otters normally have a large range, and the animals may simply have traveled well beyond where they were freed. Or, as radio-transmitter batteries wore out, otters so equipped could no longer be monitored. In any event, we do not have a continuing record of their welfare. What's more, Idaho researchers discovered that surgically implanted transmitters in otters caused pulmonary and cardiac complications resulting in two otter deaths. But despite all the setbacks, many in the wildlife community are determined to keep trying.

Because we spend many days every year in Yellowstone Park, Peggy and I have seen several dozen river otters during our travels. We've watched them in late winter along the partially frozen Lamar River and where winter's ice was retreating from the northern edges of Yellowstone Lake. We have also watched otters closer to home in the Snake River just below Jackson Lake and completely across the country in Everglades National Park. But our most intimate encounter with them came on a November morning in 1981 in Grand Teton National Park.

In the vicinity of Jackson Lake Lodge a necklace of small potholes and beaver ponds is linked by small streams, all of which eventually flow into the Snake

Winter does not greatly curtail the travels and activities of river otters. In fact, otters love to create "bobsled" runs on steep slopes.

River nearby. On passing one of the ponds that was edged with a thin crust of new ice, I saw a dark animal appear and disappear through a hole in the ice. We stopped to investigate.

During the next two hours Peggy and I watched a family of four otters catch and eat brook trout, which—judging from their brilliant colors—were then spawning in that partially frozen pond. Altogether we saw the otters capture and devour on the ice eight trout ranging in size from six inches to almost a foot. The otters soon were joined by a pair of magpies that flew down beside them to pilfer scraps and the coral eggs that squirted from the bellies of the female trout. All the while we were able to move ever closer to the action with our cameras. The incident remains doubly interesting because, until that day, I had not realized that either otters or brook trout ever traveled so far from the main channels of the Snake River. Since then, we have not found otters in these ponds or anywhere at all in the vicinity.

The North American river otter is a thick-set mammal with short legs, a neck as large as its head, inconspicuous ears, and a body that is broadest at the hips. The powerful tail is a little more than a third as long as the head and body. The hind feet are webbed. Adults weigh from 15 to 35 pounds and measure from 3 to 5 feet in length. Females are about 25 percent smaller than males. When prime, the fur appears to be dark brown, almost black, with the belly slightly lighter; the chin and throat tend to be grayish. A river otter's fur consists of a very dense undercoat with longer guard hairs on the surface.

Among all mustelids, river otters may be the most appealing to people. We tend to regard them as happy or playful because of the antics we witness in wildlife parks and zoos. We see them as clowns or scoundrels, as sleek and slippery. I've heard trout fishermen and particularly trout hatchery managers declare that there should be a bounty on otters because of their appetites for trout. Indeed, the river otter is perhaps the most graceful swimming mammal in the world. It can propel itself rapidly enough through water to catch fleeing fish, mostly just by flexing its body from its head to the tip of its tail. Otters can swim at about six miles per hour, and faster for short distances by porpoising along the surface. The species can dive to depths of at least 60 feet and remain submerged for more than four minutes. A fast man on foot could not catch one away from water. By alternately running and sliding, an otter can attain a speed of about 15 miles per hour

One of the most graceful swimming mammals, the river otter propels itself in water fast enough to overtake fish mostly by flexing its body and tail.

over ice or hard snow. An expert cross-country skier I know could not quite overtake an otter family he surprised about 200 yards from the Snake River.

But otters are not often surprised. Their vision probably isn't too good, but it is much keener underwater than on land. Their sense of smell is well-developed and their hearing is acute. Sturdy sensory whiskers on the face enable the animals to detect prey and avoid obstructions.

Much of an otter's time is spent sleeping in beaver houses, logjams, or similar shelters, especially when food is abundant and the animals do not have to spend too much time hunting. Otters of all ages are fond of play, or what appears to be play. They wrestle, slide down banks of slick mud and snow, play tag and, apparently, hide-and-seek. They also entertain themselves with rocks and driftwood.

Unfortunately, even an outdoorsman living in good otter country will see more otter sign than the otters themselves. Especially in Alaska and the northern American Midwest, the species often travels a mile or more overland between bodies of water on the same, well-defined trails year after year. Other visible signs are otter scats—twisted tufts of grass, small piles of dirt, and vegetation. Urine sprayed and scent deposited on these piles serve as identification. During the winter, otters dig elaborate tunnels and occasionally dens under the snow, with only the entrances to betray their presence.

River otters breed in spring, with mating taking place either in water or on land. Delayed implantation, a period of arrested embryonic growth, results in litters being born at any time from midwinter to summer of the following year. Two or three toothless pups are born underground, and they begin to swim and eat solid food when about two months old. Females have been known to have a litter a year for two decades. A social mustelid, the land-otter group is usually a family unit consisting of a female and her pups, with or without a male. There are also all-bachelor groups.

Otter groups seem to lack leadership. Despite their seeming sociability, otters do not hunt cooperatively or share food often. Fighting is either rare or seldom witnessed, however. River otters of Alaska also hunt and live along the edge of the sea where they dine on sea urchins and crabs, shrimp and mussels, clams, and even small octopuses. Occasionally a river otter found foraging in saltwater will be mistaken for the much larger sea otter.

**THE SEA OTTER.** Although the river otter can be seen at the seashore, the sea otter has never been known to enter freshwater. Though they occasionally come out onto coastal islands or seacoasts, sea otters normally live their entire lives in or just beside the Pacific Ocean. Their original range extended from southern California northward along all of coastal Alaska, and west past the Aleutian Islands and Russia's Kamchatka Peninsula. It then continued southward to some of the northernmost islands of Japan. Sea otters were so exploited by seafaring hunters everywhere in their range that, by 1911, all were exterminated except for small isolated pockets in remotest Alaskan coves.

Thanks to belated protection and conservation efforts to redistribute the Alaskan animals remaining, an estimated 100,000 to 125,000 now live from the Aleutian Islands to Prince William Sound. Small numbers also swim again along Russia's Commander and Kurile Islands. The species has been transplanted to unoccupied former areas in the Pribilofs, British Columbia, southeastern Alaska, Washington, and Oregon. One memorable day in 1938 a single sea otter was spotted near Carmel, California; today that area and the nearby vicinity of Monterrey and Point Reyes are the most accessible places to see a wild sea otter.

The most startling thing about sea otters is their large size. Adult males may be 5 feet long and weigh from 70 to 90 pounds—some individuals exeeding 100 pounds. Females measure 4 to 4½ feet long and average 40 to 60 pounds. A sea otter's hind feet are webbed an important feature in swimming and diving. Toes of the forefeet are short and stiff, enabling the animal to deftly handle food. On land a sea otter's gait is clumsy, and the animals can easily be outdistanced by a man. Perhaps because of their vulnerability, they are seldom found more than a few yards from the protection of water.

A sea otter's fur is possibly the finest in the world. It consists of very dense, fine underfur of inch-long fibers overlain by sparse guard hairs. The underfur varies from dark brown to nearly black. Guard hairs may be paler, sometimes silvery, giving the animal a shadowy appearance. Older animals also usually develop grayish or silvery heads. That, combined with the prominent whiskers, helps explain the sea otter's nickname, "old man of the sea."

Unlike the other seals and sea lions—which depend on heavy layers of blubber to insulate them from the cold of the Pacific Ocean—the sea otter survives by relying on air trapped in its dense fur to maintain body temperature. If the fur ever becomes soiled or matted by something such as oil from a spill or ship's bilge, insulation is lost and the animal soon dies. Thus, otters must spend much time grooming and rubbing their fur to keep it clean.

In June 1986, Peggy and I cruised among the small

Sea otters do not often rest on solid ground, where
they would be vulnerable. I photographed these
animals basking on a kelp-covered rock.

rock islands offshore from Afognak Island with Roy Randall, who owns a small lodge there. We saw many sea otters out of the water, and some of them allowed us to drift by boat to within about 40 or 50 feet before they eased into the water. On a larger rocky island fully nine miles by water from Afognak Island, Randall showed me a good example of how vulnerable sea otters might be when on land.

The remote island is also a favored haulout for Steller's sea lions. At least one female brown bear had found that by making the long swim from Afognak in near freezing water, it might capture a sea lion now and then. But apparently the sea otters, which also haul out here, were easier to catch than sea lions, and Randall showed me the remains of several that had been eaten by the bear. Roy has actually seen the bear on the island twice, once with twin cubs. The scattered bones of the otters suggest that the bear may make annual trips to prey on the otters here.

Sea otters are such powerful and graceful swimmers that they almost seem more at home in the ocean than some fish. They are capable of quickly covering great distances either above or beneath the surface. When chased—as they too often are by Alaskan commerical fishermen, many of whom despise them— they can swim like a porpoise, alternately swimming

just underwater, then arcing above the surface for air. But when really frightened and desperate to escape, sea otters dive deep and stay down to the limit of their endurance while moving away.

Mothers with young often leave pups alone on the surface while they dive far down to hunt. But if a boat approaches while they are down, they quickly rise to suddenly snatch the pup from sight.

Sea otters dive to the bottom in 5 to 250 feet of water, later returning with several items of food. On the surface they roll on their backs, place the food on their chests and, holding it with forepaws, eat it piece by piece. Clams may be cracked by striking two together. But several times off California I saw an otter retrieve a flat rock along with its food, place the rock on its belly, and strike the shellfish against it, freeing the inner morsel. Wild sea otters never dine on land.

The daily search for food consumes much of every well-fed sea otter's life. A normal feeding dive usually requires only a minute, but the animals are capable of remaining underwater for up to five minutes. To obtain the daily minimum of 10 to 20 pounds of solid nourishment needed to survive, each otter may have to eat 40 to 50 pounds of whole shellfish.

California sea otters mate at all times of the year,

Sea otters have often been observed placing a flat rock on their stomachs
to crack or crush abalone and shellfish to reach the meat inside.

During dives of from one to four minutes, this California sea otter emerged
with sea urchins and crabs that it ate while floating on its back.

Sea otters live along some of the most scenic sea coasts of North America, as these in Oregon and California. The largest otters reach 100 pounds.

and their pups may be born in any month. In Alaska most of the pups are born in the spring. As with other sea mammals, only one pup, weighing three to five pounds, is born at a time. Sea otter pups are light brown in color. For about twelve months the mother cares for her offspring, never leaving it except briefly during feeding dives. Whether traveling, sleeping, or preening, the pup either rides on its mother's back or clings to her chest as she floats on her back. A pup will weigh about 30 pounds and appear to be as large as its mother when it is finally weaned. Females rarely if ever mate as long as they have unweaned pups, which means that they produce one pup every two years.

Unlike other ocean mammals, sea otters do not migrate and may not wander far from where they are born unless that area becomes overpopulated or the food supply runs out. They are gregarious, and around Afognak Island I have seen them in groups or, more correctly, "pods" of about 100 animals. Pods up to ten times that large have also been reported. Larger breeding males will drive nonbreeders away from concentrations of females. These nonbreeders may themselves gather in bachelor areas that are usually along exposed capes and points of land where shallow water extends far offshore.

Bald eagles prey on unprotected newborn pups, and killer whales may strike sea otters of any size. But such predation is believed to be insignificant. Humans are still their most formidable enemies. Along lonely wilderness shorelines, sea otters might live from 15 to 20 years.

If they do not, it is usually because of the tremendous amount of seafood they must consume. Since too often this is the same seafood that commercial fishermen depend on, fishermen do not generally regard sea otters as the endearing, attractive creatures we watch with affection in aquaria and sealife parks. Of the fish and sea urchins, rock oysters and crabs, abalone and mussels, as well as various other mollusks and octopuses that may be included in the animal's normal diet, fishermen leave only the sea urchin ungrudgingly to the otters.

Since 1911 it has been illegal to kill sea otters in America, and their future now seems secure. Indeed, the time may not be far off when biologists determine that some small harvest could be permitted. Pressure continues from commercial fishing interests for a large harvest in the immediate future, however. They blame sea otters for the scarcity of shellfish in some areas. But studies show that the blame belongs just as much to the fishermen who have for too long overharvested certain crabs and abalone—as well as to those responsible for offshore water pollution where raw sewage pours into the coastal belt, killing the fish. In addition, future oil spills could be devastating to sea otters.

Peggy and I have spent countless pleasant hours watching and photographing sea otters from Monterrey to Kodiak. Wherever they are protected they soon become confiding and easy to approach. One thing is certain: Our northwestern coasts would be lesser places without them.

A cornered badger is an opponent to be avoided, ▷ as many a ranch dog has discovered. All but the most powerful wild predators leave badgers alone.

# CHAPTER EIGHTEEN
# THE BADGER

I've found badgers in daytime in high plains habitat, along with herds of
antelope, in central Wyoming. But the badgers were by far most active at night.

When writing about the many species of North American predators, it is difficult to separate the animals from the dedicated biologists who have studied them and discovered what makes them tick. My friends Steve and Kathy Minta are two such biologists, having pursued badgers for four years from 1982 through 1985, under both the best and worst of conditions.

Their study area for *Taxidea taxus* was the 42-square-mile National Elk Refuge in northwestern Wyoming. Under the best of conditions, in high summer and fall, their study area is famous around the world as a scenic wonder where wildlife is abundant. The cool dry weather from June through September is as stimulating as weather can be. But in winter the Elk Refuge becomes a savagely cold place where numbing winds whine, the temperature lingers below zero, and outdoor work is seldom pleasant.

Not only did the Mintas stalk and study badgers for three summers while living in a drafty sheepherder's wagon, without electricity and indoor plumbing, they also lived in that same cramped wagon throughout winters when the temperature fell to more than −20°F. Through their determination, the Mintas learned or confirmed much of what follows about this remarkable mammal. I should add that between winter field excursions and by kerosene lantern, Kathy even managed to write *The Digging Badger*, which is among the best of all children's wildlife books I have ever seen.

The badger of North America was born to dig. It excavates deep burrows in which to live. Females dig dens that may extend eight feet, in which they bear and raise young. All badgers must dig for food because other burrowing creatures are among their most important prey. They also dig to escape natural enemies and they are astonishingly fast at this, their premier skill. The Mintas recorded badgers excavating to a depth of about six feet in only a few minutes.

**BADGER HABITAT AND PHYSIQUE.** Badgers are neither the easiest nor most difficult of American predators to see. Anyone driving the back roads of the western Great Plains is more likely to see a badger at night than in daytime. I have seen badgers from Sonora and Sinaloa, Mexico, northward all the way through Montana. The species lives in deserts, plains, agricultural lands, and occasionally in mountain foothills. Its range stretches from the Pacific Ocean as far eastward as Indiana and Ohio. Although based in the Midwest during the first half of my life, I saw very few badgers there, and certainly none at all when traveling, hunting, and fishing in Wisconsin, the Badger State. Nowhere are badgers conspicuous, not even where they are most numerous.

A badger may at first seem much smaller than it is. The legs are short and stout and the creature is almost as wide as it is high, with only a short tail. Adults average about two feet long and weigh 20 to 25 pounds. A very large male might weigh as much as 30 pounds. Over most of their range the males are called boars, the females sows, and the young—born in springtime—cubs or pups.

All badgers use all four legs when digging, but the front legs are the powerful ones. They are equipped with thick, hard claws up to an inch long. The back claws are only half as long. To excavate, the digger uses its muscled front feet to break new ground as with double shovels, while the rear legs and feet clear the loose earth out of the way in what may seem to be a blur of motion and flying dirt.

By watching very closely over a long period, the Mintas concluded that the digging was ritual, not unlike a simple dance step. Every animal digs with one foot and throws earth behind and away with the opposite foot. All that happens so fast that a badger can disappear underground, leaving a pile of soil behind so rapidly that it's impossible to tell exactly how it was accomplished.

Many of the carnivores described in this book are powerful, formidable antagonists. But considering size, only the wolverine may be tougher to confront than the badger. The animals may not exactly be fearless, but they give that impression. Most badgers will hurry away to avoid any kind of trouble, but once cornered, the can cope with all but the largest, fiercest of natural enemies. I once saw a small pet badger thoroughly defeat a large, mean ranch dog that attacked it. And I've heard about badgers standing off a pack of four hunting hounds while at the same time managing to dig an escape underground.

A badger is one of those animals that can sharpen its own canine teeth just by using them. Altogether an adult animal's jaw is full of 34 teeth, the canines

Badgers are born to dig. They dig to create dens, to find food, and to escape
enemies. None of the creatures in this book are better, faster excavators.

more than an inch long. The regular, natural action of the upper canine teeth rubbing against the lower canines keeps all of these teeth extremely sharp. The jaws are hinged so that they can lock shut after biting down on prey or enemy. Thus, a badger can often deliver a serious if not fatal wound with a single bite.

A badger is difficult to mistake for any other animal sharing its range. Generally grizzled yellow-gray in appearance, this mustelid has a median white stripe that runs from the dark nose over the top of the head and to the shoulders. The badger's white cheeks and sharply contrasting dark stripes in front of the ears are more than just distinctive; the bold face pattern probably tells predators that, despite its slow gait and clumsy appearance, here is a potentially ferocious critter to be avoided. No animal that depends on hunting to survive can take a chance on being seriously injured. And the badger's facial pattern may be all the warning necessary to prevent an attack.

A badger's sense of smell, though keen, may not be as good as that of bears and some other mammals. But its hearing is superb. Any badger absolutely must hear very well or perish, because the animal must detect prey underground as well as on the surface.

The master digger has better vision after dark than during daylight. That same visual adaptation that fa-

vors sight in low light probably does not allow the animal to see in color. Instead the badger most likely sees the world in shades of gray. The Mintas' studies reveal the badger to be most active after dusk and in the cool, moist hours before dawn. Anyone looking for badgers in daylight is most apt to see them soon after daybreak.

**SCENT MARKING.** Like the other carnivores in this book, badgers use their sense of smell to communicate with others of the species. Like other mustelids, most notably skunks, badgers have scent glands from which they can emit a musky odor. It is not as foul or rank as a skunk's musk and cannot be used to spray an enemy, but is nonetheless offensive to humans. As badgers travel and hunt about their own areas, they leave notice of their passage by scent marking. Just by smelling the scent marks of another, a badger knows instantly whether it was male or female, a relative or unrelated, a local badger or a stranger, and whether it was ill or in good health.

**BIRTH AND DEVELOPMENT.** Badgers give birth in spring to softly furred young that weigh a few ounces and remain blind and helpless for about 30 days. Females dig natal dens well before they give birth. These serve as extended wombs allowing the mother to leave her pups in warm security while she goes hunting. The Mintas excavated many dens after the pups had left, and they were always astonished at the length and complexity of the tunnels, chambers, and turnaround systems, as well as at the exceptional digging ability of these subterranean architects.

As the pups grow and first emerge from the dens in late May, females are forced to hunt for a longer time and over a wider territory to feed them. In Jackson Hole country, this means prowling in search of more ground squirrels. Other predators soon realize this and are attracted to the hunting badgers, because their digging often sends squirrels scurrying in panic into the waiting mouths of the followers.

**FEEDING HABITS AND BEHAVIOR.** Badgers depend to the greatest extent on burrowing rodents such as ground squirrels, prairie dogs, gophers, and mice. Once a rodent is located, the badger quickly digs down to capture it, though it may have to dig in several spots before inflicting the final, deadly bite. Badgers can dig out several rodents in a night of hunting. A genuinely hungry badger, such as a female with pups to feed in spring, might thoroughly dig up a local area, leaving it a beehive of fresh burrows.

The Mintas selected their Jackson Hole study area because the Uinta ground squirrel—locally called a chiseler—is abundant there. Mice and chipmunks exist there too, though in smaller numbers. Additionally, the badgers are protected on the National Elk Refuge. There are a few weasels in the area, and other than badgers, coyotes are the only other terrestrial predators of importance.

During early summer when the young Uinta squirrels are still unwary and clumsy, ravens in particular follow the badgers. So do prairie falcons, and redtailed hawks soar overhead alert for a disoriented, flushed chiseler. But the most common beneficiary of the prowling badger is the coyote. I have already mentioned this remarkable relationship between coyote and badger in my chapter on coyotes; until recently such information was mostly based on old anecdotes and folklore. In fact Mexican Indians called the badger *talcoyote*—meaning "like a coyote"—because the two were seen together so often. Time and again Steve Minta was able to see this curious partnership for himself during his busy days on the National Elk Refuge.

Although as a killer of rodents the badger would seem beneficial to ranchers and farmers, the animals have often been shot and poisoned just because they exist. For that reason, I suspect they may be a lot harder to approach in daylight than might otherwise

In northwestern Wyoming, the ubiquitous Uinta ground squirrel comprises most of the badger's year-around food supply.

be the case. For example, I frequently hunted mule deer on one large ranch in Montana's Madison River valley. Here predators had never been shot or controlled, and badgers had become quite tolerant of people. I spent a good many hours sitting on the front steps of a log bunkhouse watching nearby badgers that paid very little attention to me. This is the only place where I was ever able to see young ones gamboling around the mouth of a den and, once, tumbling over and greeting an arriving adult, probably the mother.

But that experience is unusual. Normally if you get too close even when approaching a badger from downwind, the animal will disappear underground. Even if you had a shovel, strong arms, and a helper or two, you couldn't dig fast enough to catch one.

Badgers survive the long, cold winters of Jackson Hole and elsewhere by taking deep winter naps, called *torpor* by biologists. Though hibernation is similar, hibernation is a much deeper sleep during which the heart rate and body temperature decrease. Badgers wake often from torpor to go hunting on the warmer days and nights. They do not rely on fat reserves accumulated during the previous autumn alone to see them through winter. The longest period of torpor in a badger noted by the Mintas was 80 days.

Jackson Hole badgers are thus the single predator able to hunt the Uinta ground squirrel year-round. Despite the covering of snow on the ground, a badger can still locate hibernating ground squirrels deep in their winter dens through keen senses of hearing and smell. Its powerful muscles, claws, and skeletal structure allow the badger to break through the frozen soil to the doomed hibernator below. No other predator covered in this book is equipped to hunt this way in the dead of winter.

Until after being tranquilized, live-trapped badgers have proven most difficult for gloved biologists to handle. Loose skin around a badger's head and neck makes gripping badgers nearly impossible because they can turn inside their own skin and grab the handler.

After a typically severe winter, Jackson Hole suddenly comes into bloom in late spring. The land is green and incredibly beautiful. The young of most animals are born during the spring, including those of the predator species from badgers to redtailed hawks. They all take advantage of the abundant prey, especially of Uinta ground squirrels. There are so many chiselers for a short time that little competition exists among badgers, coyotes, and the raptors.

**THE PREDATOR UP CLOSE.** Kathy Minta never tired of her long hours spent observing the badger simply because the animals were never boring. Alert badgers would stare intently at her as if looking for some clue. Angry ones—particularly when faced with intruding badgers—would rise up on hind legs, noses curled back, hissing and growling. On warm spring mornings a well-fed badger might take time to roll indolently in the loose dirt around a burrow mouth,

shake, scratch, and yawn, and then briefly lie belly-up and seem to soak up the Wyoming sunshine.

But to become thoroughly acquainted with badgers, the Mintas could not rely on observations alone. They also had to line-trap them. During three years of using heavily-padded, steel leg-hold traps, they captured 66 different badgers on about 4,000 acres of the open, sagebrush study area. They also made 31 recaptures. Once in a trap any badger is a tough customer. Loose skin around the neck and shoulders makes it impossible to grab the animal safely by hand as when handling some other creatures; they can too easily turn inside their own skin and grab the handler. The Mintas thus devised a noose to slip around the neck to hold a trapped animal until they could give it a tranquilizer. Only asleep are the badgers really safe to examine.

Captured badgers reacted to live-trapping in different ways. Some would try to bury themselves, while others would simply dig out huge craters around the trap site. Many became trap-wise and, judging from tracks, wouldn't go near the devices. Retrapped badgers would try to dig in so that the expected noose could not be slipped over its head. Once asleep, every badger was aged, weighed, sexed, measured, checked for disease, and tagged. The oldest ever captured was 14 years (most do not live past six), and the heaviest was 31 pounds. Some captives were implanted with radio transmitters rather than radio-collared, because of the tendency of the collars to slide off. By using a different frequency for each animal, Minta could track and follow each badger thus equipped. He learned much about their travels and life-styles, and about why the animal has remained morphologically unchanged in the New World for roughly six million years.

Badgers are doing fairly well in the United States today, mostly because their fur has little commerical value and the use of poison baits has diminished. The reduced use of poisons is good news, not only for badgers but for all animals living in the wild and for the people who care about them.

The spotted skunk is quite widespread in North ▷ America, missing only from the Northeast and the northern Midwest.

# CHAPTER NINETEEN
# THE SKUNKS

The striped skunk is the most widely distributed of
four skunks found in the United States, all of
which have effective chemical warfare protection.

P sychologists have often noted that serious hobbies, including very strange ones, help make life more interesting and worthwhile for many of us. George F. Toland of Salina, Kansas, has a hobby stranger than most, but we may all learn something because of it. He censuses wild animals killed along highways by speeding automobiles.

For example, Toland has counted 207 skunks killed on 6,906 miles of roadways in six states in 1983. A year later he counted 250 squashed skunks along 7,422 roadway miles in six states. During 1985 he spotted 66 skunk remains along 4,061 miles in nine states. Besides showing that the nationwide toll of skunks must be very high, Toland's counts suggest that skunks, like most mammals out at night, have trouble judging the approach of vehicles. Although instantly recognizable in pictures by everyone, the only skunks most Americans ever see are those found dead on the roads. And most people couldn't care less.

"Skunk" is a name that covers four different North American animals: spotted skunk, striped skunk, hooded skunk, and hognosed skunk. The first two species are by far the most widespread and abundant. Striped skunks (*Mephitis mephitis*) range in every state and Canadian province, and in northern Mexico. The spotted skunk (*Mephitis putorius*) is missing from the northeast and northern Midwest. North of Mexico, hooded skunks (*Mephitis macroura*) are only found in southern Arizona and southwestern New Mexico. The hognose skunk (*Conepatus leuconotus*), also mainly Mexican, can be found in the Southwest as far north as southern Colorado and across the southern half of Texas. What the four share is a basic black color with the unique, contrasting white markings that distinguish one from another. They also share the ability to spray a decidedly unpleasant musk when aroused, frightened, or attacked.

Chemically the musk substance is called n-butyl mercaptan. It's an olive-green compound that includes plenty of sulfur. The liquid is carried in two scent glands or sacs, each not much larger than a pea and encased in a layer of squeezing muscles. Each is also connected to a short duct that leads to the anal tract. The duct openings are internal until the skunk stiffly erects its tail. This action pushes the tract outward like bubble gum and the twin ducts are pointed rearward at any victim like a double-barreled shotgun. The seriousness of the skunk's predicament may determine whether the barrels are fired singly or in salvo.

This seething nest of young striped skunks was found in the root system of a large, overturned dead tree. The young were still blind and helpless, but they already emitted the distinct skunk odor.

Much has been claimed about how far and how accurately a skunk can spray its jet of terrible mist. Some claim ten feet. But as far as I know, the distance has not yet been carefully researched and measured. I do know that a friend's old black Labrador retriever was accurately hit just before it was to retrieve a female striped skunk with several young. Blackie was temporarily blinded, nauseated, and probably disoriented while we scrubbed it twice with strong laundry soap in an Ohio stream. But even this double dousing, in addition to later applications of tomato juice and ammonia, did not completely eliminate the lingering aura of skunk. So make no mistake about the power of a skunk's chemical-warfare capability.

Researchers are now testing skunk musk in concentrated form as a possible bear repellent. The spray is what has enabled a sluggish, otherwise mostly defenseless mammal to survive. Fortunately, skunks probably do not spray without urgent cause. I once poked a very angry one out from beneath a log cabin fishing camp without retaliation. And Peggy and I have come upon them suddenly when hiking on woodland trails without mishap. During the 1950s I knew a professional trapper in southern Ohio, Dayton Parsons who could somehow pick up wild striped skunks without having to pay for it. It was all in the way he approached and deftly handled them.

Still, as with other wild creatures, I suggest keeping a safe distance. A skunk is probably not agitated as long as its tail is down. Blackie, the same Labrador

I mentioned earlier, did not learn a lesson from its first encounter. A few months later Blackie caught another skunk and suffered again. So did his human friends. Thus, like porcupines, rattlesnakes, and hornets, the skunk is one creature other wild animals tend to leave alone—with a couple of notable exceptions. Once in a while a hungry black bear may kill and somehow manage to eat one. And the great horned owl, which lacks a sense of smell, will take a skunk whenever it can.

**THE STRIPED SKUNK.** The striped skunk is the one most likely to show up around back porches after dark where bowls of dog food and other scraps are carelessly left outdoors overnight. About the size of a house cat—or slightly larger at its maximum of ten pounds—the species is easily recognized by the broad white area at the nape of its neck that divides into a vee at the shoulder and extends to the base of the tail. The skunk also has a narrow white stripe on its forehead. Its bushy, curved tail is usually black, though a few have white or partially white tails. And its eyes are amber in headlights or in the beam of a flashlight. But you may well smell this striped mustelid before you actually see it.

We so seldom see striped skunks because they are nocturnal, beginning each night's prowl for food shortly after sundown. Most of the road kills I enumerated at the beginning of this chapter occurred after dusk. The animals wander widely in search of mice and bird eggs, grubs and insects, berries and amphibians, as well as carrion. Some of the road mortality must be a result of feeding on other road-killed species. No itinerant skunk ever seems to stray far from water. Hunting ends at sunrise when they retire.

Dens can be ground burrows dug by other animals, or cavities deep beneath rock and wood piles. Spaces beneath buildings, usually abandoned ones, are favorite sleeping sites. I knew of one rural home in central Ohio under which at least one skunk lived for several years without the occupants ever knowing it. Several females may den together during winter, but trappers assure me the males are solitary. This species does not hibernate. In the American South they remain active throughout the year, while in the North they are out only on the warmest winter nights, otherwise sleeping in a warm den. In the most ideal habitat, striped skunk population in midsummer may be as high as one per ten acres. But a home range normally is two to five times that size.

The breeding season is spread from February in the Southeast to March in the northern Midwest. Males are polygamous. Five or six young are born blind 63 days later. I once found a litter of very young striped skunks beneath a crumbling farm outbuilding, their mother nowhere to be seen. The tiny animals already had the strong odor of skunk about them.

Striped skunks have made fair, not-too-intelligent pets when surgicallly de-scented at an early age, but they really belong in the wild rather than in a home. Besides, even the most attractive of them when young, grow into fat, irritable adults.

**THE SPOTTED SKUNK.** The spotted skunk is the most striking, and the easiest to identify of its family. It has a white spot on the forehead and under each ear, in addition to four other broken white stripes along the neck, back, and sides. The white tip on its tail is immediately noticeable. Overall the pelage might be described as having a harlequin pattern.

Spotted skunks are nocturnal and are similar in most of their other habits to striped skunks, though their home and hunting ranges are larger. Rabies have been found in both, and any abnormally friendly or aggressive animals should be carefully avoided. Eastern and midwestern spotted skunks grow larger than their western cousins. Males weigh from 1 to 2½ pounds apiece while females weigh about a third less on average. Especially when judged by the number of mice and rats they kill around barnyards and farming operations, skunks are allies of man. Even so, many are trapped for the fur, though it is usually not in great demand compared to that of other mustelids.

**HOODED AND HOGNOSED SKUNKS.** Hooded skunks have two general color patterns, with intermediate variations, and are the blackest of the four American species. In one pattern, the entire back and top of tail is white or blends into white; the other pattern includes two thin white side stripes that are not always easy to see. The hair on the neck is often ruffed, and the tail, which is as long as the body, is the longest of all the skunks. This skunk adds small snakes to the usual fare of rodents, eggs, insects, and berries.

There is no mistaking the hognosed or *rooster* skunk, which might weigh as much as 6 pounds. It is bicolored, mainly all white on top and all black beneath and on the long, piglike face and snout. Its tail is all white, and its fur is short, coarse, and well suited to its warm, dry environment. The solitary hognosed skunk dens in cliffs, rockpiles, and canyons.

The only hognosed skunk I ever met, in northern Mexico, behaved exactly as others I have heard and read about. It crouched under a thorny bush, growled drily, and slapped the hard ground vigorously with both front feet. After studying the animal for a minute I hiked away, leaving well enough alone.

# PART FIVE
# THE RACCOON AND RELATIVES

The following oft-told story may or may not have happened, but it is certainly possible and even probable. It seems that one damp night soon after they were ferried ashore and left alone at Plymouth Rock, the Pilgrims huddled around a driftwood bonfire and pondered their dilemma. There they sat in a hostile new world, in wilderness an ocean away from friends and everything familiar. Then suddenly a cur dog, which had also made the voyage from England, barked at a nearby tree.

Some of the men investigated and found that the dog had treed a masked animal—a raccoon. They quickly captured it, skinned it, and cooked the animal over the fire, then thanked God for the unexpected bounty. With that roasted raccoon came the first realization that their new home in Massachusetts, as

Seldom are raccoons found far from water's edge, be it freshwater lakes and
streams or tidal marshes. More than most predators, raccoons have learned
to cope with civilization everywhere.

well as the entire continent, contained an almost miraculous abundance of wild game on which they might survive.

Speaking of survivors, the raccoon, *Procyon lotor*, is a master among masters. Like the coyote, it is more abundant and widespread now than that evening when it was "discovered" at Plymouth Rock. It would be hard to nominate a more intelligent and resourceful mammal.

**RACCOON.** To date, raccoons have taken civilization in stride. They easily survived the Roaring Twenties, without the least dip in population, when every college student in America had to have a raccoon coat. Even the Davy Crockett craze of the mid-1950s, when many small boys wore coonskin caps, had no lasting effect on raccoon numbers. (Incidentally raccoon tails that were worth 25 cents a pound in 1954 jumped to $5 a pound a year later.) Indeed, hunting, trapping, and the annual heavy highway kill are probably necessary to spare this species from the ravages of hunger and disease that come from overcrowding.

Wild raccoons are not extensive travelers as wild carnivores go. As long as the living is easy, adults may spend a lifetime in an area ranging from a few acres to four or five square miles. By contrast, young ones driven out of the family home range have been known from tag recoveries to wander as far as 30 miles seeking territory unoccupied by other raccoons. One young coon had covered 165 miles in its search.

The home territory of a raccoon can be along a meandering stream in Indiana, on the edge of a public campground in central Florida, or in the center of an eastern city or suburb. An indefinite but substantial number now survive in sewer systems and around refuse dumps. Raccoons have also become familiar figures in affluent suburbs where residents regularly feed them, probably unwisely.

About the size of a small dog, this medium-size member of the family Procyonidae weighs from about 12 to 35 pounds as an adult. Northern raccoons average much heavier than those in the south. There are occasional reports of males weighing as much as 50 pounds, but those are doubtful. To my knowledge, the largest individual reliably recorded was a 48-pounder captured by hunters near Milan, Ohio, in 1950. That exceedingly fat specimen may have escaped from a game farm. A raccoon's fur is pepper-and-salt colored, but the best identifying marks are the black mask over the eyes and the alternating rings of yellowish-white and black on the tail. It is from those features that the common nicknames of *masked bandit* and *ringtail* were derived.

Raccoons have five toes on each foot, with non-retractile claws, and they walk on the entire foot. The jaw contains 40 sharp teeth. The animals grow from 2 to over 3 feet long, and can climb and swim as well as they can walk on solid ground.

Raccoons do not hibernate. But when bitter winds whistle through the bare branches of northern forests, they sleep in hollow trees or other dens for extended periods, with tails curled over their sharp muzzles and black button noses. Following every warm spell there are paw prints in the snow and mud softened by the sun. In February, most of those tracks may be made by males searching for females in estrus.

Raccoons are prolific. Mating takes place in midwinter in the Southeast, and in late February or March in the North. Many females breed before they are a full year old. After a gestation period of about two months, two to seven young (but most often four) are born, each weighing 2 to 2½ ounces. Their eyes open three weeks later. They begin following their mother at about two months and leave or are driven away in the fall. It is during that autumn dispersal that many of the young animals are taken by hunters. Raccoons that manage to survive a hunting season or two become extremely tough to catch, even for the best hunters with the best hounds. Individual coons have also gained considerable fame locally for their remarkable ability to escape hunters season after season.

Young raccoons make up the bulk of those run over on the highways. Naturalist Kenneth Chambers once found three young coons that had been killed by a car on a New York country road. Pausing to examine them, he discovered a fourth curled up fast asleep on the berm nearby, apparently unwilling to leave its siblings. Figuring the fourth might also be killed, Chambers carried it by the scruff of the neck, kicking and squealing, across a weed field and far away for its own safety.

Raccoons can be as unerring as bears in finding their way back to a point of origin. Some years ago, Jim Donahoe, a game warden in Ohio's Champaign County, transported a semi-pet raccoon to a point eight miles away after it became a nuisance around the house where it lived. But the animal quickly found its way back, even passing through the town of Urbana to reach its destination. Taking it much farther away didn't deter the animal, either. The second time it needed three months to return, but did so nonetheless.

The raccoon is another carnivore that is chiefly nocturnal, yet often hunts during daylight as well. Daytime hunting occurs especially in coastal marshes during low tides that expose edible morsels. The species now exists in all of the American states except Hawaii and Alaska, in Mexico, as well as in a narrow strip across southern Canada. Mostly a woodland species, the 'coon is rarely found far from lakes, rivers, swamps, or seashores.

An entire book could be written about the antics and intelligence of raccoons. A biographer of Davy Crockett claimed that the Tennessee woodsman and politician could grin a coon out of a tree. When I was young I hunted raccoons regularly at night with a backwoods Kentuckian, Cy Flaugher, who kept a pack of black-and-tan coonhounds for that purpose, which was his lifelong passion. He also always tried to keep a live coon or two on hand to train his younger dogs. Cy obtained these coons by sending me, equipped with leather glove and burlap sack, into a tree where an animal had been chased. Once in awhile I managed to return to earth safely, if clawed and bitten, with the coon in the poke. It was never easy. Cy then trained his pups by dragging the wet raccoon in its sack across the forest floor and encouraging his allies to follow its scent.

The only trouble was that Cy, like a lot of other people, could never keep a coon for very long in a cage. Somehow it always managed to find a way out, at times by strength and prying alone, but more often by manipulating the latch or lock with deft hands.

Raccoons seem able to get in and out of anywhere. I knew of one that could unsnap the chain from the dog collar around its neck.

During the 1950s biologists in the eastern half of the United States noticed a drastic decline in the number of wood ducks. The blame was correctly placed on the cutting down of too many mature forests; wood ducks require the hollows of old, large trees for nesting places. Many states subsequently began building and erecting wooden nesting boxes in wet woodlands where the woodies normally raised their young. The only trouble was that too many raccoons found the boxes and either used them for their own dens or simply ate the eggs and ducklings they found inside.

To replace the flawed wooden models, Ohio officials designed a "rocket" wood-duck nest box that they believed would be predator-proof. These were cylindrical, fashioned of welded metal, with cone-shaped roofs (hence the rocket name). But of the first 35 rockets erected on the Mercer County Wildlife Refuge, biologist Dave LaRoche found raccoons in 22 of them. Later some of the boxes were smeared with a substance supposed to repel raccoons and other mammals, but that, too, was a failure. Ten coons moved

I photographed this visiting raccoon one morning while cooking breakfast outdoors in a Washington campground. Many fellow campers were familiar with this regular freeloader.

into these and raised a total of 12 young. Now 30 years later, no one has yet been able to devise an artifical wood-duck nesting box guaranteed not to accommodate raccoons.

City life invariably causes unusual problems and bizarre incidents for coons. In Lakewood, Ohio, just before Christmas, a neighborhood dog chased a local raccoon onto a roof which it refused to leave, even when carloads of firemen arrived with ladders and nets. When pressed too closely, the flustered coon darted into the chimney and didn't stop running or falling until it emerged in the basement two floors below. There it escaped through an open door, but not before overturning a decorated Christmas tree and leaving a thick trail of soot behind. At least the homeowners did not need to hire a chimney sweep that year.

Familiarity sometimes breeds contempt, and an old boar coon doesn't always run away when a feisty family dog comes nipping after it. One such raccoon drowned a beribboned French poodle in its master's swimming pool in a suburb outside Atlanta.

At Shawnigan Lake, British Columbia, J. T. Perry stored several crates of apples under his back porch for later use. He did not realize the fruit was rotting until late one night when he heard a terrible commotion beyond the back door. Looking out, Perry found four raccoons having a massive binge for themselves, gorging on the fermented apples and—he swears—staggering and behaving in a drunken manner. Perry drove them away with a shotgun and got rid of the apples the next morning. But the following night the raccoons returned looking for more.

While on a deer stand in northern Michigan I've watched coons feeding on the apples that had fallen in abandoned orchards and were beginning to soften and rot among the yellow leaves of fall. The animals will eat almost anything biodegradable, but the usual diet more often includes such aquatic prey as frogs, crayfish, and small snakes—any of which they dunk or seem to wash at streamside before eating. They will never pass up bird eggs or small nestlings, but I do not believe they are as good mousers as is often written.

A single raccoon can make a shambles of a patch of sweet corn just as it fully ripens, ready for human pleasure. In the old days when every small family farm included a chicken coop, a passing raccoon might kill a good many of the birds during a single night's foray. City dwellers who feed raccoons from back porches

The area around Anhinga Trail in Everglades National Park is a good place
to find foraging raccoons very early in the morning and at sunset.

know how much they relish canned cat and dog foods, or scraps of Kentucky fried chicken. A Pennsylvania coon learned how to reach into a pheasant hatchery at night and pull off the birds' heads. America's original masked bandits have become far greater nuisances around many campgrounds than black bears have ever been. The fortunate thing is that they do not grow as large as the bears.

The species is not without natural enemies in addition to humans and their hounds. Any of the carnivores from black bears to cougars, jaguars, and fishers will kill a raccoon for food if they can catch it. The list of potential predators of younger raccoons is even longer and would certainly include coyotes, bobcats, and mink. A baby raccoon was once found in the stomach of a Florida diamondback rattlesnake. Because they frequent water, raccoons must be wary of still another predator in the Southeast. While photographing wood storks early one morning, I saw an alligator catch and devour a raccoon in Florida's Corkscrew Swamp. That probably is not uncommon. In fact raccoons would likely be far more destructive of many heron and egret rookeries in the South, pillaging both the eggs and nestlings, if they did not have to swim across alligator-infested swamps to reach them.

## COATIS AND RINGTAIL CATS.

Two relatives of the raccoon live today in the southwestern corner of the United States and in Mexico south to Panama: the coati *Nasua narica,* and the ringtail or *miner's* cat, *Bassariscus astutus.* Neither is as abundant or as familiar to most Americans as the raccoon.

Although fairly adaptable, coatis are basically creatures of tropical woodland and montane forest. They're found in southern Arizona, New Mexico, and along the Rio Grande River in Texas. Although mobile and adaptable enough to cross deserts and other unsuitable habitat, coatis do not expand farther north because of the lack of food and cover, as well as the increasing cold. Their population within the United States reached a peak in the late 1950s, but dropped suddenly, apparently because of canine distemper or a similar disease.

Coatis are opportunistic omnivores wherever they live, eating whatever fruits, invertebrates, and small vertebrates are easily obtained. From the anteater-like snout and the barber-pole tail usually held aloft, the coati could be, as Ernest Thompson Seton wrote years ago, a cross between a raccoon and a monkey with a little dash of pig thrown in.

Most of what we know about coatis comes from studies made in Panama. Females and immature males

Prowling around Corkscrew Swamp, Florida, this careless young raccoon fell prey to an alligator. This may happen fairly often in the Southeast.

The coati is the opportunistic southwestern cousin ▷ of the raccoon. A true omnivore, it feeds on fruits, invertebrates, or small creatures.

Female coatis and immature males travel in groups that are formidable enough to drive off most of their would-be predators.

live and travel in troops of 10 to 20, but the older, adult males are usually solitary. In the American Southwest and northern Mexico, the home ranges of the troops are much larger than in the tropics, and the animals may be semi-nomadic. They have been known to breed in forests above 3,000 feet in the Animas Mountains of New Mexico and farther west in the Baboquivari Mountains in Arizona.

I have never come across a troop of coatis, but have been assured that it is an interesting experience. They hiss, spit, and lash their tails. They usually shun trouble, but stick together in a brawl and, with their sharp teeth, could punish larger animals.

In late spring, eleven weeks after mating, pregnant coatis seek rocky niches high in lonely canyons to bear litters of from four to six young. These have darker coats than the rust-colored fur of adults. The promiscuous males take no part in raising young, and

in Latin America are known as *gatos solos*—lone cats.

There was a time when it was fairly fashionable to keep coatis as housepets, and young ones could even be bought from mail order houses for $49.95 apiece in 1958. But buyers soon realized that their once-lovable young pets quickly grew to 15 or 20 pound maturity. At that size, a coati was difficult to housebreak, very active, and could demolish a household. Understandably, the fad soon ended.

The raccoon's other relative in America, the ringtail cat or *cacomistle*, is another carnivore seldom spotted in the wild except at night in the species' limited range. Especially from spring through late fall, ringtails spend daytimes dozing in dens cooler than the outside. At dusk they set out to seek breakfast, which may be a desert rodent or a bird, eggs or insects, small reptiles or cactus fruit. Naturalist Olaus Murie once found a cave in New Mexico where ringtails dined on bats.

Ringtail cats are easily distinguished by their large dark eyes and foxlike faces, and by their black tails that are as long as their bodies and usually have eight white rings. The animals weigh two to almost three pounds when fully grown.

Many woodsmen and foresters who know them best believe that ringtails can control mice and wood rats better than the most efficient house cat. In fact it is a blessing to have a cacomistle take up residence around a home, cabin, or barnyard in America's southwestern mountains. Years ago a ranger, on leaving his forest-fire lookout, left this note for his successor: "Take care of the ringtail you'll soon meet here. Feed it scraps to keep it around, and I guarantee you'll never be bothered by rats."

Several years later the retired ranger revisited his old cabin and asked about his pet. The new ranger pointed to a cardboard box hidden in a dark corner. It contained four sleeping ringtails.

"I haven't seen a rodent around here since I came," the new ranger said.

# PART SIX
# PHOTOGRAPHING THE PREDATORS

Stalking America's predators to photograph or simply watch them is a fascinating and compelling pastime. The animals themselves and their life-styles are among the most interesting on earth. Photographing them only adds to that interest and, depending on the species, ranges from fairly easy to nearly impossible. The elusive large cats are in the latter category. Among the easiest are the coyotes and bears of some national parks, or the raccoons, foxes, and skunks that freeload around backyards of surburban homes.

But somewhere, somehow, there are ways to capture in the camera viewfinder all the animals we've covered in this book. In my earlier volumes in this series—*Erwin Bauer's Deer in Their World*, *Bear in Their World*, and *Horned and Antlered Game*—I considered in detail the subject of general wildlife photography. But because photographing these animals is somewhat unique, I will confine this chapter to the different techniques that are required for predators, which include more extensive use of blinds, baiting, and calling. But first let's consider basic photographic equipment.

**CAMERAS.** A skilled outdoorsperson who knows his subjects well may be able to take wildlife pictures with almost any kind of camera. But without doubt the best camera for shooting wild mammal pictures today, especially of carnivores, is the 35mm single lens reflex. There are many 35SLR models in a wide range of prices. The 35SLR is the nearly perfect choice because it is compact, sturdy, reliable, fast in operation, and—compared to other designs—relatively light in weight. The greatest advantage of the 35SLR is that the subject is viewed, and focused on, directly through the lens. You freeze on film exactly what you see in the viewfinder the instant you expose the film.

My 35SLRs function very well in extremely hot and dusty conditions, as well as in cold, damp weather. The cameras feel comfortable in my hands and seem almost an extension of me. As I type this (and knock on wood), more than a year has passed since any of my four camera bodies or other equipment have required repairs that I could not make myself. And my gear gets heavy use. One more point: I never attempt anything more than minor adjustments

Better and better cameras are introduced so frequently that it is almost impossible to keep up with developments. Some are so virtually foolproof and fully automatic that the user merely loads film, aims, and shoots. All this is well and good, and certainly accounts for high-quality pictures much of the time. But although the newest technology is built into the cameras we use, Peggy and I still prefer the manual focus but do take note of the built-in exposure meter, which we use as a guide. Photographing furtive creatures, often in motion in poor or changing light, is a lot different from other photography where the action is rarely so fast and where the subject itself can be controlled. Perhaps I am reluctant to relinquish control over my work to microchips that cannot do their figuring fast enough in the field.

To photograph any wildlife, and especially preda-

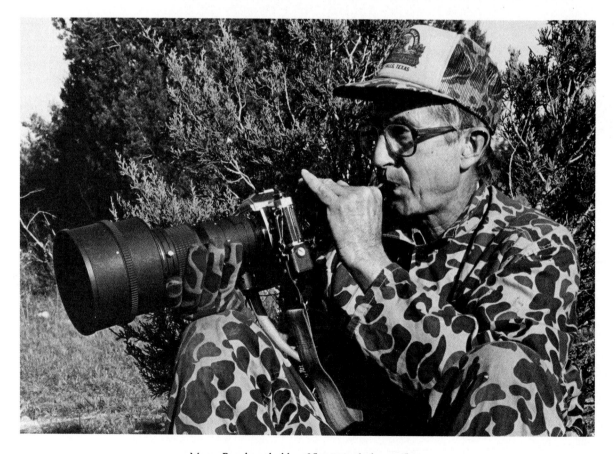

Murry Burnham holds a 35mm single-lens-reflex
(35 SLR) camera with telephoto lens as he
imitates the sounds of a distressed rabbit, in an
effort to attract predators.

238

The combination of a firm tripod, a smooth ball head, and a camera with
motor drive is what makes action photos such as these possible.

tors, the ideal 35SLR should be part of a system of interchangeable lenses that includes fast telephoto lenses. Most of the time the photographer cannot get close enough to his subject without a telephoto lens. As with cameras, better, sharper, and faster telephotos are being introduced all the time.

Photographing together as we normally do, Peggy and I always have three or four replacement camera bodies available in case of malfunction. Of course the bodies are compatible and easily interchangeable with one another and with all our lenses. They also have viewfinders on which information is shown around the edges of the frame. That way we can note the shutter speed and lens opening while we focus and compose a picture without lifting our eyes from the viewfinder. This feature is doubly convenient when light conditions are constantly changing or the subject is moving.

All our cameras are also equipped with motor drives that advance film. We feel these are well worth the added cost. Following any action—such as the running wolves pictured in this book—is smoother with a motor drive, which leaves you free to concentrate through the viewfinder without the distraction of manually

advancing the film. Motor drives can also rewind film, but doing that puts a heavy drain on the batteries. We have become so adept at rewinding film that we can do it nearly as quickly as if it were done with the motor.

Occasionally the metallic noise of the motor drive has made nervous predators even edgier and has completely spooked a few. That's a definite disadvantage. But most predators are not unduly disturbed, and often the strange sound of the first film advance has caused the animal to look up, bright-eyed and alert, for the second (and better) frame a split-second later.

The latest feature offered in 35SLR camera systems is *autofocusing,* which will eventually make obsolete all 35SLRs that do not have it. Already autofocusing can improve some wildlife and action phtography by arresting moving subjects in sharper focus than is possible manually. When the shutter release button is depressed halfway, an invisible beam of infrared light bounces from the subject to a sensor in the center of the viewfinder. That activates a tiny motor that instantly focuses the lens at that distance. Most of the autofocus SLRs available at this writing also have an autofocus lock. That lets you first focus on the subject, then lock the focus and recompose in the frame for a better photo where the subject is not in dead center.

There are also serious drawbacks to autofocus, especially when the action is fleeting. The autofocus cameras Peggy and I have examined still do not focus sharply on a subject coming toward the camera when they come into very close range. Worse still is that the infrared beam causes the camera to focus on whatever it strikes first, for example on the brush that partially obscures the subject standing just behind it. This is inconvenient for a photographer aiming at a bobcat or a marten in a thicket. The photographer can change the camera to manual focus in such situations, but that takes precious time.

I have written it before and I can't repeat the advice often enough: Half the secret of shooting successfully in the wild is thoroughly understanding how to use the equipment quickly and accurately. Practice and more practice will make you better, if not actually perfect. You don't have to know all about the mechanics and technology inside the camera—just as a skillful automobile driver doesn't have to be a good mechanic or automotive engineer.

**LENSES.** Because photography is our life work, Peggy and I always keep handy more lenses than we usually need. Ninety percent of the photos in this book were made with just three telephoto lenses: an 80–200mm f/2.8 zoom, or variable focus; a 300mm f/2.8; and a 200–400mm f/4 zoom. Of the three, the two zoom lenses may well be as sharp as similar lenses with fixed focal length. In fact there are many inexpensive zoom lenses available today that are extremely sharp and excellent for wildlife photography. The zoom's flexibility and lighter weight—less than the combined weight and bulk of several fixed-focal-length lenses—is welcome in nature photography.

Too many photography writers have treated tele-converters, sometimes called *tele-extenders,* with the same disdain as zoom lenses. But for shooting the predators as well as other wild creatures, the 1.4× extenders we always carry in a pocket have often proved invaluable. For example the 1.4×, easily inserted between lens and camera, instantly makes the 300mm lens a 420mm lens—and a lynx almost half again as big in your viewfinder and on your photo frame. The same extender converts our 200–400 f/4

As described in accompanying text, Peggy and I use just three of the many lenses here for nearly all of our predator photography. The large lenses in the back row are all Nikons; left to right they are 600mm f/4, 80–200mm zoom f/3.5, 400mm f./3.5, 200–400mm f/4, 300mm f/2.8. The small lenses and accessories in front, left to right, include the following: Nikkor 55mm f/3.5, Vivitar 5600 flash, Nikkor 35–70mm f/3.5, Vivitar 70–210mm f/3.5. The two small lenslike accessories at the extreme right are tele-extenders.

zoom into a 280-560mm f/5.6 zoom. The loss of one f/stop is no problem in bright light. As with any lens, some teleconverters are not as sharp as others, so shop carefully.

**TRIPODS.** Photographing predators differs from photographing their prey in one important respect: With predators, it is more a matter of attracting the subjects into telephoto-lens range than striking out in search of them. There are numerous exceptions to that, but most often you're playing the waiting game.

Because the photographer is not moving as much when photographing wild predators, a suitable tripod becomes more important than ever. A tripod is a necessity when photographing from any kind of fixed blind. Unfortunately, selecting the right one at the start can be as bewildering as choosing the "best" of all 35SLRs.

Because you may not have to carry it often or far, a heavier, sturdier tripod can be used for photographing predators. We've settled on one with telescoping tubular metal legs that easily adjusts from eye level to very low height so it is as easy to use sitting down or kneeling as when standing up. The unit also easily and quickly adjusts for uneven or steeply sloping ground. Keep in mind all of these requirements if you are shopping for the ideal tripod. And be sure to set the tripod up and try it out before making your purchase.

The head connecting the camera and telephoto lens to the tripod is at least as important as the tripod itself. I've never found an acceptable tripod head to be inexpensive or even modestly priced. Look for a large ball head that allows you to aim, up and down, right or left, so fluidly that you can concentrate entirely on focusing and composition. It should also hold the camera and lens pointed exactly where you want it, without slipping even slightly or requiring great force to screw it tight. Finally, the tripod head should include a quick-release feature, as opposed to a screw-in mount. The quick-release system, with adapter plates on all our telephoto lenses, permits rapid mounting or changing cameras on the tripod heads.

**ELECTRONIC FLASH.** I have often written that we avoid using flash for photographing wildlife sub-

With 35SLR and 600mm lens mounted on a sturdy tripod, I am here equipped to photograph bears at fairly long range—a precaution that helps reduce the risk of being attacked. For frame-filling images it is usually necessary to approach much closer to most other predators.

The large knob for this tripod head allows one-hand loosening and locking of the ball. The top knob allows quick insertion and release of the camera, fitted with adapter plates, eliminating the need (on many tripods) to laboriously turn the camera onto a threaded tripod bolt.

jects. Most birds and animals accept flash without becoming alarmed, so that is not my reason. Rather, I dislike the studio effect that results; it eliminates the wild look for which I strive so diligently. But because many predators are nocturnal or rarely emerge from shadows and dismal light, a flash is often needed.

With most 35SLR cameras, electronic-flash photography is complicated because the flash can only be synchronized with shutter speeds of 1/60 or 1/80th second. Yet the flash itself may occur in less than 1/1000th second. In darkness that is no problem. But when a flash picture is snapped in dull or dappled daylight, there may be a ghost image in the photograph because the animal was moving before, during, and after the brief flash itself, while the camera shutter remained open for 1/60th or 1/80th second. The ghost image does not appear if the subject is motionless while the shutter is open. Unfortunately, you cannot count on your subject's remaining still at this crucial moment.

There are so many portable flash units on the market today that a whole volume would be required to discuss them adequately. The ideal light for our type of wildlife shooting is the small battery-powered portable that either attaches directly to the camera or is connected to it by a short arm. Most of the best portable flashes are either fully automatic or can be operated manually. I usually use the autoflash on which

I use a Fresnel lens mounted in front of my flash unit to extend the range of the flash. This is useful when shooting shy subjects in forest shade.

I simply set the f/stop, based on film speed, and the range of the flash. When I squeeze the shutter, light from the small flash reflects from the subject back to a sensor on the flash unit. That determines the duration of the flash, usually from 1/1000 to 1/30,000th of a second. The exposure is usually correct, unless there is a foreground object (leaves, tree branches, or grass, for example) that intercepts the light reflection to the sensor.

A drawback to using a flash unit mounted directly on the camera is that the subject may appear flat and shadowless, with unnatural red or green eyes. You can avoid that by aiming the flash from one side and above the level of the camera. Try handholding the flash unit or using a flash connector arm.

Another flash factor: The effective range of the light from many portable flash units is too short. It may not throw light far enough to reach a shy creature crouched some distance away. The range of some units can be extended by using a Fresnel lens in front of the flash. This lens concentrates the light and sends it farther than it would ordinarily reach.

This point cannot be overemphasized: Before doing any serious flash photography of wildlife *outdoors*, thoroughly test your camera-lens-film-flash combination on some convenient, stationery subject *out in the field*. Most of the advice, recommendations, and guide numbers given in flash-unit manuals are made for *indoor* shooting where there are reflective surfaces such as walls and ceilings. Outdoors there is no such ambient light. The only way you can possibly learn what your own unit will do is through extensive testing. Trial and error wil be your most effective learning method.

**PREDATOR LOCALES.** Except for the core areas of large cities, there are few locales in the United States where it is impossible to photograph one or more of the predators covered in this book. But for most species, the best places by far are the national parks of the United States and Canada, or other large wildlife sanctuaries where hunting is banned and all creatures have had years to become accustomed to people and their activities.

During my first visits to Denali National Park, Alaska, in the 1950s, I didn't see any of the wolves there. But on later visits we've spotted more and more of them, each time a little nearer to the single trans-park roadway. On several occasions in 1986 Peggy and I found wolves hunting just beside that road, oblivious to tourist buses passing in clouds of dust. Twice we were able to photograph them barely 25 feet from our van, something that would have been inconceivable not so long ago.

**BAITING, BLINDS, AND CALLING.** Several of the predators can best be brought to bay and then photographed by pursuing them with trained hounds as a hunter would. In Chapter 4 is a photo of a jaguar driven to swim in a swamp in Colombia by a pack of dogs. Another jaguar is treed.

I'm embarrassed to admit how many tedious hours I have spent, often with mosquitos buzzing about my ears, waiting in a tree blind for a jaguar to find the live bait tethered just below. Still, baiting is often the most effective technique to lure almost all of the meat-eaters within telephoto range. Raccoons, skunks, and bears are the easiest of all to bribe with food; but because of the possible consequences, I do not recommend baiting bears anywhere except far from human dwelling and activity.

Baiting also means using blinds most of the time. And blinds fall into two general categories: permanent, fixed blinds, and portable blinds. Which one you should use depends on the situation and the subject.

Where a photographer maintains a bait or feeding station on a permanent basis, such as near a waterhole in the Southwest, a more-or-less permanent blind is the choice. Good blinds can be constructed of scrap items or of natural materials gathered near, but not right on, the spot because you don't want to disturb the natural setting more than necessary. When building such a blind, the first consideration should be the sunlight. Or if you will use the blind after dark, also consider where you will place artificial lights. The second consideration is comfort.

Since you are already going to the trouble of erecting a blind, why not make it roomy enough so that you do not have to wait for long periods in a stooped or cramped position? If the blind is not fairly windproof, use plastic or old canvas sheets draped along the walls to blunt persistent drafts. A photographer cannot really concentrate on his work if he is cold for long spells. In hot climates make provision for adequate ventilation. It is also a good idea to pad the floor to eliminate or muffle the sounds of booted feet. Be certain there is sufficient space at the proper level to cover the entire baited area with a telephoto lens, but don't use too large an opening. The main handicap with most fixed blinds is that they cannot be conveniently moved, which always seems to become necessary at some point.

A serious wildlife photographer is wise to keep a portable blind handy. The perfect one has not yet been invented. By that I mean a blind light enough to carry with ease anywhere, which can be quickly erected and will withstand the elements and especially high winds. This blind should also be roomy enough

Here we've created a spacious blind from driftwood for the photographing of brown bears.

This portable easy-to-pitch Rue blind was used to shoot some of the photos in this book.

inside and have zippered camera ports at various levels all around. We have found the best of several now on the market to be the Ultimate Blind, manufactured by professional photographer Leonard Lee Rue III, of Blairstown, New Jersey. It weighs only 9½ pounds and can be pitched in a minute or two.

For camera-hunting of the large horned and antlered mammals, I rarely try to conceal my presence at all and in fact believe the animals can be better approached if they can watch and see me at all times. But concealment is essential for most predator photography. Instead of using a blind, it may be possible to take fullest advantage of whatever natural cover and concealment you find. Under those circumstances

Here I'm equipped with camera gear and portable blind—everything necessary to "shoot" such carnivores as coyotes, bobcats, foxes, and raccoons.

the photographer must also wear camouflage clothing from head to foot.

This is an ideal point at which to reintroduce my old friend Murry Burnham, of Marble Falls, Texas. Murry ranks high among the best woodsmen I have ever met, and he certainly must be the greatest wild-game caller living. Except for grizzly bears, otters and wolves, Murry has successfully called into close range at least one of every predator described in this book. He once called a cougar in Mexico across the Rio Grande to the Texas side of the river. (Burnham is convinced that the only reason the wolf is not on his list is that he has not yet had sufficient opportunity to test his techniques on them.) Using his own manufactured calls or personal tapes of his calling, Burnham has coaxed by the thousands such species as coyotes and bobcats into range close enough for photography. Some of these have approached near enough to touch, and so convincing has been his calling that at least one coyote and one fox have nipped his arms. One day Murry called up seven coyotes for me in a single morning, and that is nowhere near his record. A good many of the photographs in this book are either a result of Murry's calling, his equipment, his advice, or all three.

A long telephoto lens not only makes grizzly-bear photography possible, it also makes it safer, though never entirely safe.

Burnham correctly believes that much of his success comes from being absolutely inconspicuous in the field. He moves slowly and almost as quietly as any wild cat. He also wears outer garments of camouflage cloth which make him almost invisible. That includes gloves and—because he wears glasses—a cap with camouflage netting that covers his face, though some photographers might prefer to use the camouflage makeup now available.

Whenever possible Burnham thoroughly scouts a new area, looking for fresh predator sign as well as for the best places in which to sit hidden before he begins to call. The following morning before daybreak, having done his preliminary work, Murry sits quietly in one of the spots he found earlier and begins to call.

For all of the predators, Burnham uses a small plastic or wooden mouth call that imitates the cry of an injured jackrabbit in the jaws of a meat-eater. Apparently this pitiful sound has immediate appeal to all but the wiliest predators. At the very least it excites their curiosity to investigate.

At first Murry blows at the top of his lungs, putting as much pleading and terror into the sound as he can, so that it can be heard far away. Then he pauses for a few mintues before blowing again, this time more

Murry Burnham demonstrates the use of calling and a stuffed jackrabbit to attract small predators, as described in the text. The decoy proves fascinating to carnivores.

Without hesitation (and without spotting the photographer), a coyote attacks the jackrabbit decoy. (Murry Burnham photo)

In lieu of a decoy, another Burnham trick is to hang a rabbit or coyote hide on a bush in full view next to the amplifier. Many predators cannot resist rushing in for a closer look.

softly. At this point, often a coyote or bobcat will at least become visible somewhere nearby. After another pause, Burnham blows again, this time very faintly, pleading, as if the victim is about to expire. As I have often seen myself, this faint cry is more than many carnivores can resist. Not only is this a technique for getting good animal pictures, but it is also a sport to be avoided by any with weak hearts.

Of course just blowing a predator call does not guarantee instant success. It takes time and subtle shading or muting of the sounds to entice suspicious or older animals into range. But even inexpert mouth calling will produce some camera subjects.

A modern alternative to mouth calling and years of practice is a cassette player/amplifier that can be carried into the field. These have some notable advantages of their own. Tapes of anything from squalling cottontails to wounded javelinas can be broadcast. To make the performance even more effective, Burnham has assembled a complete outfit, which includes a speaker that, by cable and reel, can be placed 50 feet away from the person using the cassette caller. It is amazing how often the predators will zero in on the sound and scarcely notice the person crouched and busy photographing nearby.

Hounds treed this young cougar long enough for
me to shoot close-up photos. Cougars may also
respond to predator calls.

For information on mouth calls, cassette/players,
amplifiers, tapes, and camouflage clothing, write to
Burnham Brothers, Box 669, Marble Falls, Texas
78654.

One cool winter morning in 1983 in south Texas
brush country, Murry promised to show me a new
trick he had been perfecting, not only to attract but
to keep coyotes in camera range for extended photography. One problem with calling alone is that a
wary predator will not hang around anywhere very
long after discovering it has made a mistake.

Murry's trick involved the use of a Texas jackrabbit
mounted in a somewhat lifelike pose on a board with
oversize ears spread in a vee. Just before sunup he
placed the decoy in the center of a clearing. To one
of the jackrabbit's ears, using fine monofilament fishing line, he tied a ball of white fluff—actually the
tail of a cottontail rabbit. The fluff moved or swung
gently with the slightest breath of wind, giving life
to the whole decoy. As soon as we had enough light
for photography, my friend turned on the cassette
player with the speaker placed very near the stuffed
jackrabbit. The two of us crouched hopefully nearby.

There was no agonizing wait this time. A coyote
charged out of the brush and stopped suddenly, def-

The pine marten is one of the small predators that
can be easy to attract into close camera range with
bait. Anything from a dead squirrel to a spoonful
of raspberry preserves can work.

initely perplexed, before it focused on the phony hare.
I thought it would turn and run away, but just then
the ball of fluff moved ever so slightly. The coyote
sprang forward, struck the dummy, and grimly hung
on to the back of its rigid neck for several seconds
before letting go and disappearing.

I was so surprised by what I saw that I didn't get
any good pictures of it. But since that revealing
morning, Murry's battered decoy in conjunction with
the calling has helped me fool countless other shy
coyotes. One of the coyote photos he took is shown
herewith.

Thus the old proverb, "There is more than one
way to skin a cat," applies here. There are more ways
than I have mentioned to photograph the most interesting, most mysterious mammals in North America. As an addicted wildlife photographer, I try to
discover and improve on more and more of them.

◁ The combination of a low afternoon sun and an
inquisitive gray wolf, helped me get this shot.
Wolves are among the most difficult creatures to
photograph, owing to their fear of man and
normally remote habitats.

# INDEX